Unfare Solutions

Transport, Development and Sustainability

Series editor: David Banister, Professor of Transport Planning, University College London

Barry Ubbels
Marcus Enoch
Stephen Potter
Peter Nijkamp

Unfare Solutions
Local earmarked charges to fund public transport

Spon Press
Taylor & Francis Group

LONDON AND NEW YORK

First published 2004 by Spon Press
11 New Fetter Lane, London EC4P 4EE
Simultaneously published in the USA and Canada
by Spon Press, 29 West 35th Street, New York, NY 10001

Spon Press is an imprint of the Taylor & Francis Group

Typeset in Sabon and Imago by PNR Design, Oxfordshire
Printed and bound in Great Britain by The Cromwell Press, Trowbridge, Wiltshire

British Library Cataloguing in Publication Data
A catalogue record for this book is available from the British Library

Library of Congress Cataloging in Publication Data

Ubbels, Barry
    Unfare solutions: local earmarked charges to fund public transport/Barry Ubbels (et al).
        p. cm. – (Transport, development and sustainability)
    Includes bibliographical references and index.
    ISBN 0-415-32712-1
        1. Transportation–Taxation. 2. Transportation–Finance. 3. User charges.
    4. Infrastructure (Economics)–Finance. I. Ubbels, Barry. II. Series

    HE196.9 .U54 2004
    388.4'042–dc22   2003018540
M K ISBN 0-415-32712-1

Contents

In almost all developed economies, and increasingly in developing ones as well, transport has become a major problem for policy-makers, particularly in urban areas. The traditional response to increasing car use, that of road building, is now hard to justify economically, socially, and (above all) environmentally. Instead, the aim of most transport policy-makers is now the management of travel demand, a key element of which has been the development of an attractive public transport system as an alternative to car travel. But for this to work, major sustained investment is required. Such investment is difficult when public spending is already stretched to fund improvements to other parts of the public sector. Added to this, there are general economic and competitive pressures. In a deregulating and liberalized global economy, it is hard to sustain a high taxation regime.

One strategic approach has been for governments to cut costs through privatization, efficiency savings, or a combination of the two, and this has become an increasingly trodden path over recent years. The results are patchy and have not always yielded sufficient investment funds or enhanced services to the level needed to meet policy needs. Less common are examples of local transport authorities raising money specifically to pay for improvements to public transport through dedicated local charges and/or taxes. This latter response may even be combined with privatization with, for example, the new income stream paying the annual charges to a private operator for upgrading a public transport system.

As well as simply raising money for public transport development, such new sources of finance can themselves be tools of mobility management. The most obvious examples are road user pricing and parking charges. The issue of new sources of finance for public transport investment and operations is one that exists whatever form of ownership or regulation model is adopted, and the links between these financing mechanisms and transport policy are increasingly important.

This book examines such sources of local earmarked finance. Its purpose is to explore the linkages between these charging mechanisms and modern transport policy and finance. It seeks to identify and present cases of creative ways of funding public transport (or mass transit). Traditional financing mechanisms for public transport remain important and even indispensable to maintain or enhance market share and quality, but complementary finance based on innovative local funding initiatives can offer a more competitive position for public transport. The present study aims to discuss the principles of and experiences with such alternative funding mechanisms. A distinctive feature of the book is that it brings together in a systematic way world-wide experience of earmarked urban public transport funding mechanisms. Its novelty is to be found in both the interesting – and sometimes fascinating – cases and in the theoretical framework provided in order to position these cases.

The book is intended to offer new information on the practical relevance of such new funding systems to transport operators, urban and regional policy-makers and funding institutions. The theoretical framework also serves to offer new scientific insights to the transport research community.

The 'struggle' between private and public transport may be eased and relaxed, if public transport is able to come up with creative solutions for complex budget deficit issues. The 'Asphalt Nation' (Kay, 1997) is not a necessity, but its emergence also depends critically on non-traditional initiatives in the public transport sector. Seen from this perspective, the present publication is not only informative, but also missionary in nature.

The first Chapter looks at the environmental, economic, and social challenges of transport trends, while sources of finance for policy interventions are the focus of Chapter 2. The following three Chapters then detail a number of local earmarked financing measures. These have been grouped into a series of categories (such as 'beneficiary pays', 'polluter pays', and 'spreading the burden') that relate to the policy context and fiscal basis of their design (see Table 0.1).

The measures used range from the fairly prosaic, such as employment taxes, property taxes, and developer levies, to somewhat more bizarre revenue raising schemes. In certain locales, it may be encouraging to learn that undertaking oft-frowned upon activities such as drinking, gambling, smoking, driving, flying, and shopping, are actually contributing to improving public transport services. This is thanks to taxes on beer (Birmingham, Alabama), lottery tickets (Arizona), cigarettes (in Oregan), parking/fuel/cars/vehicle parts (Stansted Airport/Florida/Chicago/Seattle among many others), aircraft landing fees (JFK Airport, New York) and shop sales (Atlanta).

It may be comforting to think that the local taxation of such 'vices' is helping fund public transport, but is this anything other than a local source of easy money? Can taxing vices really solve the transport crisis? Crucial strategic questions such as these are addressed in the final two Chapters of the book.

Table 0.1 Summary of the categories of local earmarked taxes and charges analysed in this book

Category	Methods	Description	Features
Beneficiary pays	*Employee/ employer taxes*	• usually a local charge per employee • sometimes banded with highest payments in areas of best public transport • sometimes relief for employers who provide public transport support to staff	• a simple, low cost and practical mechanism that can be effective in providing a reliable and substantial fund • possibility of companies/public to locate outside public transport accessible areas • acceptability initially difficult, but where the transport system is seen as problematic, businesses may be keen to help address the problem
	Property taxes	• tax upon property in areas of public transport • 'user pays' concept: intended to capture some of the rise in property values generated by public transport • usually earmarked business tax • often used to pay loans/bonds	• a simple, low cost and practical mechanism that can be effective in providing a reliable and substantial fund • widely used in United States and subject to voter approval in North America
	Developer levies	• can be applied in a variety of ways, including by private developers • often linked to planning permission	• a transferable scheme with varying practicality over the various identified cases • usually small scale implementation but high acceptability
Polluter pays	*Parking charges and fines*	• the use of parking charges or fines to fund public transport • applied by both private and public authorities • makes use of existing powers	• a simple, low cost and practical (transparent) mechanism that may provide substantial funds • acceptable and transferable system • linked to both transport and environmental policy
	Road space charges	• includes tolls, congestion and road user charges • may require new powers • can raise large sums	• a flexible and transparent system with a large potential to support public transport (because of political interest and large revenues) • (public) acceptability may be problematic • linked to both transport and environmental policy
	Local motor taxes	• includes local levy on fuel and excise taxes	• a large source of revenue, depending on travel patterns • transferability depends on existing tax structure • acceptable as fuel taxes are common practice; voter approval required in North America • linked to both transport and environmental policy
Spreading the burden	*Consumption taxes*	• local taxes on a variety of consumption goods and services • may be a general goods/services tax or on a particular good (e.g. beer or gambling) • used extensively in the United States	• transferability might be difficult as these schemes are dependent on North American circumstances • tend to be acceptable as voting is necessary, but significant community outreach has to be completed • significant source of revenue although influenced by external factors
	Cross utility financing	• where multi-utility companies provide a subsidy to public transport from their other operations	• a dedicated source of funding with low costs • not really practical anymore for EU countries due to new legislation
Combination	*Miscellaneous*	• rest category including airport landing charges and student fees to pay for public transport	• a simple system to collect and easy to understand • effective as it provides a specific service which might not have run otherwise • might be problematic to transfer due to local specific circumstances

Chapter 1

Transport in a sustainable society

The mobility explosion

Since the early 1950s all developed countries have witnessed a 'mobility explosion'. Indeed, across the fifteen countries in the European Union (EU-15), overall passenger transport use (in cars, buses, coaches, trams, trains and aeroplanes) rose by 121% between 1970 and 1996. This translates into an increase in the average distance travelled by each EU citizen per day from 16.5 km to 35 km over the same period. Transport demand across the EU was calculated as 4700 billion passenger-kilometres in 1996 (EC, 1999).

The majority of this increase is due to a rise in car use, although air transport is experiencing the fastest increase of all, albeit from a lower level than for cars. Over the period 1970–1996 car use increased by 136%, with the modal share increasing from 74% of passenger-kilometres in 1970 to 79% in 1996. This has been facilitated by increased road capacity, with income and population growth viewed as the major driving forces behind increasing vehicle ownership and use (Marshall et al., 1997; Marshall and Banister, 2000). In the EU-15, there was a 34% increase in the number of vehicles owned between 1985 and 1995, with the number of cars on EU-15 roads growing from 60.77 million to 165.54 million, an average growth rate of just less than 4% a year. Thus, by 1996, there were 444 cars per 1000 EU-15 inhabitants (EC, 1999). The OECD (1995) predicted that this would increase by a further 50% between 1995 and 2020, bringing vehicle ownership levels to more than 600 per 1000 people in many EU-15 countries.

Such growth is frightening enough, but at the moment 80% of the 550 million vehicles (including 400 million cars) registered world-wide are owned by the richest 15% of people living in the 'mainly developed' and industrialized OECD countries. Unsurprisingly therefore, the number of vehicles and associated traffic levels are growing much faster in developing countries than in the developed world. Two-thirds of the rise in vehicles is forecast to occur in non-OECD nations particularly in Eastern Europe and Asia. If historic rates are maintained, the global vehicle population will exceed one billion by 2020 (Potter, 2000).

Meanwhile bus and coach ridership either grew or remained stable across the fifteen European Union countries between 1970 and 1996, and with

(a)

(b)

(c)

Figure 1.1 Traffic congestion in (a) London, (b) Paris and (c) Port Louis, Mauritius. Traffic growth and transport problems are now a global phenomenon

an overall increase of 39%.[1] However, the crucial point is that despite this increase, the average modal share of bus and coach in the EU fell from 12.5% to only 7.8%, once again emphasizing the scale of the growth in car use (EC, 1999). Interestingly, even in the heavily car dependent United States, bus use was 5% higher in 1997 than it had been 20 years previously, although patronage dropped in between these dates (APTA, 1999).

Travel is now a major part of virtually everyone's budget and lifestyle. Generally this is viewed as part of our increasing standard of living, but there does exist a counter argument that a lot of this additional travel is viewed less positively – the need to travel so much can be seen more as an unwelcome burden on our lives. Unfortunately, the mobility which provides us with the freedom to drive away at weekends to a rural retreat also locks us into the daily commuting grind, forces us to ferry our children everywhere, and leads to us to battle with several thousand other cars to reach the out-of-town superstore or shopping mall. The benefits and burdens of our high mobility lifestyle tend to be rather intertwined. Transport, and high car use in particular, are increasingly recognized as one of the major political issues facing local and national governments across the world. Consequences include congestion, delays, inconvenience and stress for the individual, quite apart from the cumulative negative impact mass car use has on the economy, the environment, and on communities.

Mobility megatrends and the transport crisis

It is not just that transport is becoming an increasingly important part of our economy and society. It is also an increasingly complex activity, and this increasing complexity itself results in rebound effects that further increase transport dependence. Transport of both people and goods has come to involve interlinked, multi-modal and geographically connected networks, which in themselves stimulate even more transport dependence. The separation of home and workplace began many years ago, but we are now moving to a situation where the radius of action of our lives is an ever widening one (Van Doren, 1992). The rise in car use has resulted in local shops being replaced by the out-of-town supermarket, the retail warehouse and shopping mall. Hospitals and schools are getting bigger and more remote, journeys to work are increasing in length and, more and more, we find ourselves having to make a whole series of long and complex mixed journeys just to carry out 'everyday' activities.

Increased mobility has accelerated the long-term trend to more dispersed and lower density settlement patterns. This pattern has been further reinforced as long-term demographic trends have lowered densities of occupation. This has been due to a reduction in the numbers of children born; more young adults setting up home on their own rather than living with their parents; the increase in divorces and separations; and longer life expectancy leading to an increasing proportion of one- or two-person 'elderly' households.

This trend towards dispersed and lightly occupied settlement patterns itself has major transport implications. Although well suited to the car, such places have very difficult operating conditions for public transport, which suffers a double blow. Firstly, it is used less because people have better access to cars, and secondly, the remaining demand consists of lighter loads spread over a large number of routes. Furthermore, reduced population catchments lead to the decline of local facilities. Overall, access and mobility becomes polarized. Those with a car readily available have unparalleled freedom of travel. Those who do not have a car (particularly children, the old, women and people with disabilities) find themselves increasingly isolated.

Although not entirely a product of more cars and better roads, increased mobility has played a major enabling role in stimulating these economic and lifestyle trends. The Information Technology era seems to be taking such trends further forward, creating the constellation of a network society, with a merger of local and global connectivity patterns (Castells, 1996). Today, to many in the developed world, a shopping trip might equally consist of a few minutes walk down the road, or an Internet order placed on the other side of the globe.

Our modern network economy has ultimately become very mobility intensive; it particularly favours (if not requires) car use and longer trips, and stacks the odds against public transport. Local access achieved by walking and cycling becomes less and less viable as everyday trips shift further away. Each improvement in transport is more than absorbed by an increase in its use. The idea that faster and more comfortable cars, better roads, high-speed trains, or new air services save time is nonsense. The individual improvements may appear to do so, but in practice these improvements are soon accommodated into a higher level of overall mobility. The concept of the 'constant time budget' (Zahavi, 1973) reflects this situation. It is demonstrated by empirical data from household travel surveys throughout the world. The British National Travel Survey, for example, notes that in the 20 years 1976–1996 the amount of time spent travelling actually increased slightly from an average of 330 to 358 hours per person per year. In the same period the number of journeys undertaken hardly changed, but the distance travelled rose by nearly 40% (Potter, 2000).

The way in which improvements in transport are simply part of the process that increases transport intensity has important implications. This is particularly so regarding the environmental impact of transport. If more transport efficiency at the micro level (fuel efficiency, lower cost, higher speeds, network access) results not in a cut in resources used per unit of GDP, but in an increase, then the final effect will be environmentally degrading. It is only at the overall system level that the environmental (or the economic and societal) impacts of transport can be evaluated (Potter, 2000). This major implication of transport's megatrends has yet to be fully appreciated by transport policy-makers.

Transport and the environment

Transport produces detrimental environmental impacts at a number of levels. This is a growing issue that affects all developed and developing economies alike. According to Elsom (1996) poor urban air quality threatens the health and well-being of about one-half of the world's urban population, and notes that it is likely to get worse. This is because of rapidly increasing urban populations, unchecked urban and industrial expansion, and the phenomenal surge in the number and use of motor vehicles. Despite California's stringent emission standards for cars, air quality for the 14 million inhabitants of the Los Angeles basin fails to meet Federal standards on 130 days each year (albeit an improvement on the 226 days in 1988). In Mexico City, the situation is even worse, and the smog can be so severe that industrial plants are ordered to cut production by 50%–75% and schoolchildren are given the month off. In China, the most common cause of death is now respiratory illness brought on by air pollution (from all sectors not just transport), and some cities are so thickly covered with air pollution that they are not visible on satellite photographs.

A useful hierarchy of environmental impacts featured in the Dutch National Environmental Policy Plan (Dutch Ministry of Housing, Physical Planning and Environment, 1990). Transport's environmental impacts include:

◆ local – noise, smell, air quality, health effect, particulates, volatile organic compounds, carbon monoxide, ozone;
◆ regional – waste disposal, land use, land take of infrastructure;
◆ continental – acid rain, nitrogen oxides, sulphur dioxide; and
◆ global – climate change, ozone depletion, carbon dioxide, ozone, CFCs.

In the nineteenth century, mankind's influence on the biosphere was local and immediate. In recent years, however, the impacts of society on the environment have become increasingly large-scale, long-term and in some cases almost irreversible. Examples of these new kinds of issues are acidic deposition, the Antarctic stratospheric ozone hole, deforestation, desertification, and losses of genetic resources. One particular type of human activity that impinges on the environment in a multiplicity of ways is transport. Transport plays an important role in a country's environmental performance

Table 1.1 Total transport emissions as % of total emissions (mid 1990s)

	Nitrogen oxides (NO_x)	Carbon dioxide (CO_2)	Sulphur oxides (SO_x)
North America	51	31	4
OECD Europe	60	24	5
EU-15	63	26	7
OECD States	52	27	5

Source: OECD, 1999

and the sustainability of its development. Table 1.1 provides information on emissions of key pollutants by the transport sector.

Local impacts: air quality

Local air quality issues, particularly the health effects of car emissions, increasingly feature in the popular media as well as in the technical press. A European Commission study (Bjerrgaard et al., 1996) noted that deaths, hospitalization, work sick leave and other health effects attributable to traffic pollution amounts to at least 0.4% of GDP, with estimates as high as 3% being suggested in some cases. Emissions from the transport sector are the largest single source, responsible for about 60% of the total, amounting to 0.38% of the country's gross domestic product (GDP) – very close to Bjerrgaard's EU estimate.

In major cities of developing countries the air quality is already worse than that in cities of industrial countries, despite lower levels of vehicle ownership. Road traffic is not the only source of air pollution, but it is the primary source of some important categories of pollutants (such as carbon monoxide and nitrogen oxides). These emissions damage health, especially of persons living or working in the open air. In Mexico City, for example, high particulate levels contribute to an estimated 12,500 deaths a year (Serageldin, 1993). In the UK, the 1998 Transport Policy White Paper (DETR, 1998a), noted that 'up to 24,000 vulnerable people are estimated to die prematurely each year (in the UK), and a similar number are admitted to hospital, because of exposure to air pollution, much of which is due to road traffic'. The death toll is over six times the number killed in road accidents.

At this point it is salient to compare the relative lack of attention to the deaths and illness caused by vehicle pollution with the strong media coverage around more remote environmental health hazards. For example, no fatalities have so far been proved to have been caused by genetically modified crops, yet the possible risk receives far more media attention than the deaths of 24,000 people each year in the UK as a result of the effects of air pollution.

Many developed nations have programmes at both national and local levels to bring the concentration of these pollutants to a 'safe' level. At the national

Figure 1.2 'Cut Pollution' publicity poster, Manchester, UK

and international level there have been major initiatives to develop cleaner fuel formulations together with 'clean-up' technologies for vehicle exhausts. In the EU all leaded petrol was finally phased out in 2000 (and earlier in the US), although it is still used in developing countries. The sulphur levels in both diesel and petrol have also been lowered. Air quality improvements were the aim of the mandatory introduction of catalytic converters on all new cars in the US from the early 1980s and in the EU from 1992. These substantially reduced the amount of carbon monoxide and nitrogen oxides emitted. In the United States, catalytic converters have been mandatory for 20 years but traffic growth has negated their beneficial impacts upon urban air quality. By the 1990s, this led to regulations seeking further improvements, including legislation in California and other US States requiring the production of 'Zero-Emission' and 'Low-Emission' Vehicles. Thus, rather than simply cleaning existing petrol and diesel, the aim has been to replace it with 'alternative fuels'. Although the US legislation (subsequently diluted) stimulated the development of battery electric vehicles, other alternative fuels have proved more practical, even if not resulting in the total elimination of on-street emissions. These include vehicles powered by Liquified Petroleum Gas, Compressed Natural Gas, manufactured alcohol fuels and petrol or diesel-electric hybrids.

Despite the cleaning of existing vehicle technologies, and the gradual emergence of new ones, the sheer volume of vehicles and the projected growth in traffic makes it increasingly hard to hit 'safe' air quality standards. This realization is behind the pressure for more radically 'clean' technologies, but also expresses itself in the acceptance, as noted earlier in this chapter, of the need to also manage the number of vehicles. Air quality concerns are particularly acute in larger towns and cities and it is here that traffic management for local environmental and health concerns is of particular importance.

Local impacts: wider health issues

The worsening air quality caused by emissions from transport sources reflects a growing concern for the health effects of our transport systems. The landmark Buchanan Report (Ministry of Transport, 1963) discussed three links. Most prominent were road traffic fatalities and injuries because of the many thousands of lives lost or permanently impaired. Secondly, noise pollution,[2] partly because it is measurable but also because some effects are tangible, for example lost sleep. Thirdly, air pollution which, although the Report noted that 'engine fumes do not yet rank as a major cause of atmospheric pollution', did refer to the carcinogenic properties of fumes and also smog causing eye and throat irritants.

While all these continue to be of concern, increasing attention is being paid to links between impairment of lung function and motor traffic emissions. Today there is a growing consensus that the alarming rise in child asthma is related to the ability of emissions to lower tolerance thresholds. Most recently PM10s (particles of less than 10 μm in diameter) associated with diesel have heightened concerns that such pollution may cause cancers.

It is now beginning to be recognized that transport's impact on health also involves more subtle and cumulative processes than the above distinct issues suggest. These include behavioural and lifestyle changes such as reductions in independent mobility as traffic levels rise. For the elderly this may include withdrawal from street life and loss of social support networks, with the associated increased health risks. Over several decades Hillman's work has charted such effects, particularly with regards to children (Hillman, 1993). Loss of independent mobility may damage children's emotional and physical development. This is due to the decline of safe and accessible space for play and exercise, including the school journey. This both reduces children's ability to explore and learn about their environment and contributes to increasingly sedentary lifestyles with consequent concerns about fitness and heart health (Cale and Almond, 1992). Importantly it is known that sedentary children are likely to become sedentary adults, perpetuating poor health into adulthood. Such changes are in part, if not wholly, responses to the hostile, polluted, noisy and dangerous street environment of which motor traffic is the prime cause.

Thus the 'systems' health impacts of transport extend well beyond the direct effects of pollutants emitted or casualties inflicted. So, for example, parents drive their children to school and forbid them to play outside because of fears of road traffic accidents. As a result, many children do not get enough exercise.

Regional, continental and global environmental impacts

The targets and technical measures that are emerging to reduce transport's adverse impact upon local air quality still leave the more strategic, regional, continental and global environmental issues unaddressed. An area of particular concern is the contribution of carbon dioxide (CO_2) emissions to global warming. Catalytic converters only marginally increase fuel use and CO_2 emissions, while even radical responses to the air quality issue, such as electric vehicles, largely transfer pollution and emissions of CO_2 from the street to the power station, although some improvements in efficiency are possible.

In order to address this, the 1992 Climate Change Treaty was drawn up and signed by most developed nations, providing an international obligation for them to stabilize CO_2 emissions at 1990 levels by the end of 2000. This was in turn superseded by the 1997 Kyoto Protocol and 2001 Bonn Accord. EU Member States have pledged to reduce emissions of greenhouse gases by 8% from 1990 levels in the period 2008–2012, and have entered into a complicated set of national emission targets that distribute growth or reduction in carbon emissions among the fifteen nations. Interestingly, under a special provision, the EU and the Member States are allowed to fulfil their commitments jointly through a burden sharing agreement (IEA, 2001).

As the fastest-growing source of CO_2 emissions, the transport sector is by far the greatest cause for concern. While air transport is growing fast, it is road transport that is currently the major problem area. Across the European Union, carbon dioxide emissions from transport increased from 733.8 MtC (megatonnes of carbon) in 1990 to 825.4 MtC in 1996 – an average annual growth of 1.9% over the period – increasing its share of total energy output from 20% to 26% (EC, 1999). Another EC study has demonstrated that, without some form of policy intervention, CO_2 from the transport sector would rise by 40% between 1990 and 2010 (EC, 1997).

The Intergovernmental Panel on Climate Change (Houghton *et al.*, 1990) estimated that to halt the net growth of CO_2 in the atmosphere, and so limit the effects of global warming, emissions must be reduced worldwide by at least 60%. To allow for a more modest target by developing countries, the industrially developed world should be seeking a greater than 60% cut.

Until now, it has been clear that EU Member States have largely avoided trying to make cuts in the transport sector's carbon dioxide contribution, preferring instead to target the less politically challenging domestic, commercial and industrial sectors. If, however, the IPCC target is ever to be a realistic aspiration, this will have to change. In 1990, the total energy consumed across the EU-15 was around 3670 MtC, with transport accounting for around 735 MtC. Reducing overall emissions by 60% would require the total figure to drop to 1470 MtC. If transport's emissions grew by the 40% suggested in EC (1997) to 2010 (to around 1030 MtC), then to meet such a 'sustainability' target would require cutting CO_2 emissions from the domestic, industry and commercial sectors to a mere 440 MtC – a cut of nearly 90%. Such figures border on the ludicrous and show that, although failing to get transport to take its fair share in cutting CO_2 emissions may have worked in the short term, it simply worsens the twenty-first century's environmental crisis.

As this book is concentrating upon funding sources to develop public transport services, the issue of the scope for modal shift to public transport to cut CO_2 emissions requires some consideration. In a related study (Potter *et al.*, 2001), we made an estimate of the reduction CO_2 emissions produced by modal shift from single-occupancy car commuting to public transport. The estimate was based on detailed data from the British National Travel Survey, but is likely to be comparable to elsewhere in Europe. According to the National Travel Survey, the average car commuting by people employed full or part-time was 5,933 km per year as a car driver. A litre of petrol produces about 2.4 kg of CO_2 and a litre of diesel about 2.7 kg. A figure of 2.5 kg per litre would represent an average for all cars, allowing for the petrol/diesel mix in the car stock. The average UK fuel economy is 9 litres per 100 km, although the driving conditions for commuting trips might well involve a poorer fuel economy than the average.

If the average fuel economy were taken, then each single car occupancy commuting trip produces about 1.3 tonnes of CO_2 emissions per annum. However, if local urban transport policy leads to a reduction in CO_2 from car travel and an increase in the use of public transport, then the increase in CO_2 emissions from the latter should be taken into account. A review and survey of the primary life cycle fuel consumption of a wide variety of vehicles (reported in Potter, 2000) indicate that public transport in peak hours used less than 20% of the energy consumed by a single occupancy car. Thus for every peak hour car trip diverted to public transport, the *net* CO_2 saved would be 80% of the *gross* cut in CO_2 from the car, which is just over 1 tonne of CO_2 per annum.

Transport casualties and safety

The growing awareness of the direct and indirect health impacts of transport, considered above, has been coupled with heightened concerns about the continuing high level of deaths and injuries from road accidents. Deaths from road traffic incidents in the EU-15 have been reduced. Over the period 1970– 1996, they declined by an impressive 48%, from 73,556 deaths (221 per million inhabitants, or 301 per 1000 million passenger-kilometres) to 41,806 fatalities (112 per million inhabitants or 149 per 1000 million passenger-kilometres). This reduction in road deaths has been due to a number of factors. These include improved road design, tougher legislation on drink driving, higher vehicle safety standards (crashworthiness and design of vehicle exterior for pedestrians protection), introduction of speed limits, stricter rules on the driving times of professional drivers, reduced truck load capacities, and improved monitoring of the roadworthiness of vehicles.

But, transport casualties remain an important issue. Over 1.7 million people in the EU-15 are still injured each year, and the annual costs of transport accidents were estimated at $200bn (€224bn) in 1998 (ECMT, 1998).[3] This is quite apart from the human suffering, and the fact that traffic accidents remain the primary cause of death for persons under 40 years old. A fatal road death represents an average loss of 40 years, compared with 10.5 years for cancer and 9.7 years for cardio-vascular illnesses (EC, 1999).

Equity

Despite this massive growth in car use, the proportion of so-called 'transportation disadvantaged', i.e. those who do not have a car available

Figure 1.3 Another road casualty statistic

at the time they want to make a trip, still amounts to more than half the EU population. Even where families can afford a car though, it is often the case that one adult uses this to travel to work, meaning that the rest of the family must do without. However, there are not just economic barriers to using cars. There are also many people who cannot use a car, because they are physically or mentally unable to do so, or are too young to gain a driving licence. In addition, there is a sizeable number of (especially elderly) people who prefer not to drive even if they can (Black, 1995).

These groups, the poor, the handicapped, the young, and the elderly suffer increasing mobility problems as car use increases, especially as land use patterns become increasingly dispersed and as public transport services become less economic to run (and are therefore curtailed) as a consequence (Pucher and Lefèvre, 1996).

Economy

Congestion is an increasing cost both to economies and society. Car use has outstripped the growth in road capacity and so congestion has worsened (Marshall *et al.*, 1997). This is particularly so in cities where little new infrastructure can be built. Overall, this has led to average speeds in cities to decline by approximately 5% per decade (EFTE, 1994) with increased costs to businesses and society generally due to the resultant time delays. Black (1995) also attributed a proportion of accidents and psychological strain as being due to congestion.

Across the EU-15, the total cost of transport externalities, such as congestion, accidents and environmental impacts, is substantial. Eurostat (1997) estimated the cost of congestion as amounting to around 2% of the EU-15 GDP, which is about €175bn.[4] ECMT (1998) estimated road traffic accidents to cost 2.5% (c.€220bn), uncovered infrastructure 0.15% (c.€13bn), local air pollution 0.6% (c.€53bn), noise 0.4% (c.€25bn) and climate change 0.5% (c.€44bn) of EU-15 GDP each year. In addition, on average 30–40% of land in each European city is taken to satisfy the need for parking spaces, petrol stations and roads, while increased reliance on the car has led to the severance of communities, etc. Because more and more people are using cars, the problems are set to worsen, which in turn makes the public transport less viable, and walking and cycling even less pleasant for those who are denied access to a car.

Challenges for policy-makers

Until the 1990s, most nations' response to traffic growth involved 'demand-led' policies whereby government investment is used to increase road capacity roughly in line with traffic growth. This 'demand-led' or 'predict and provide' approach should not be confused with the fact that some countries invested heavily in high-quality public transport systems for their cities. To a large extent this simply reflected the transport intensification viewpoint that provided major new roads. More and more, transport capacity was the predominant regime approach, although in big cities the emphasis might be a little more on this being in metro systems rather than urban motorways.

The credibility of the traditional 'demand-led' transport approach gradually collapsed in the late 1980s and 1990s. This came about by a mixture of factors. Firstly there was the failure of the traditional road building response to achieve its prime aim of reducing traffic congestion and transport costs. The roots of this were in the rebound and feedback effects noted above. An example is a series of studies (Goodwin *et al.*, 1991 and Goodwin, 1994) that indicated that Britain could not physically, economically or socially accommodate the 1989 Department of Transport forecasts of a 110% increase in traffic (Department of Transport, 1989). Even a road building programme of an inconceivable vastness would fail to stop congestion getting worse. As a policy response, road building will always fail; transport demand management is the only direction possible for transport policy at all levels.[5]

But, to the simple failure of the predominant policy regime was added a new, crucial, factor: the growing awareness of motor traffic's global environmental impacts.

On the ground, the transition to demand management policies has proved to be politically fraught and, across Europe, *mobility management* has occurred only in isolated cases. For example, in the Netherlands proposals to introduce a road user charging scheme for the densely populated Randstad area in the west of the country have been repeatedly postponed and recently abandoned due to political opposition. In Germany, the strength of the car lobby has ensured that even a policy as innocuous as introducing speed limits on the *Autobahn* motorway network has been resisted tooth and nail for many years. In the United Kingdom, witness the limited interest of local authorities in adopting road user charging and work place parking levy legislation. This reluctance is despite the assurance that monies raised from the charges will be earmarked to pay for improvements to their wider transport infrastructure and services.

There is an extremely important lesson here, and one that has been put aside in the wake of transports' environmental concerns. This is that mobility management is not simply needed to address environmental concerns; it is required anyway to address transports' economic and social impacts. It is simply physically and economically impossible to continue to meet the historical and projected demands for road traffic. However, it is environmental considerations that, through the 1990s and into the twenty-first century, have come to be a key driving force in the transport policy debate.

But, while increased reliance on the car has caused all sorts of problems for society as a whole, on an individual level there are numerous benefits that can be derived from the use of cars. This returns us to the strategic discussion at the beginning of this chapter on transport megatrends and the nature of the transport policy crisis. We appear to be trapped in a high transport intensity system that is set to require more and more resources be devoted to travelling to achieve the same output, while generating a whole range of unsustainable impacts – at the environmental, economic and social level.

The depth and extent of the car dependency regime was commented upon perceptively by Freund and Martin in their 1993 book *The Ecology of the Automobile*:

The sensual, erotic, or irrational well springs of the auto mobility cannot be ignored. The pleasure, as well as the convenience that auto driving provides is a boon to many people. However, what is needed is a transport system that allows people to find pleasure in many ways of travel. New policies must be as non-punitive as possible in discouraging auto use, and must develop seductive, as well as affordable and efficient alternatives to the auto.

This view is shared by Belk (1995), who noted:

An automobile is not just a transportation vehicle, but a vehicle for fantasy fun, prestige, power, pollution, carnage, sex, mobility, connection, alienation, aggression, achievement, and a host of cultural changes it brings in its wake.

The crux of the problem is that the benefits of car use are very evident to individuals, whereas the problems are more diffuse, hit others rather than the car user, with some impinging on future rather than current generations. Car use causes changes that cannot easily be perceived and which are accommodated on a piecemeal basis. This means that any individual action produces no obvious personal benefit. For example, one mother letting her children walk to school will not improve their health or safety, so long as no similar action is taken by other parents. Worse, transport policy is not an area favoured by politicians, certainly compared to, for example, education and health. For things to improve significantly, large sums of money are required, and significant improvements take many years to achieve. Transport projects can also be controversial and, even where they are generally accepted, schemes under construction tend to generate much hostility at a local level due to the disruption involved.

This unequal conflict between choosing immediate and tangible personal benefit over a delayed and far less visible cost to society is behind many of the difficulties faced when addressing the transport crisis.

Public transport

Given this situation, to achieve a public transport system of the quality needed not only to attract car users, but to provide an alternative to the whole dispersed, car-oriented regime we now have would require substantial policy involvement and funding from society as a whole. Further, it would involve a transformed system of regulation, pricing and subsidies, not only at the national and transnational level, but also a focused portfolio of local and regional initiatives. This raises many questions, such as:

◆   What are the spatial, institutional, social, technological and financial barriers faced in tackling the private car and enhancing the attraction of public transport?

◆   What is the relationship between national (and EU/Transnational) policy goals and local/regional pricing initiatives?

◆   Is it possible to provide a financially feasible portfolio of incentives and disincentives so as to favour the development of an environmentally-friendly transport regime at the local/regional level?

◆   Do local/regional authorities have sufficient power and capabilities to develop a system of local charges and taxes?

◆   Can the development of an environmentally-friendly transport regime at the local/regional level be compatible with supporting socio-economic equity and economic development?

◆   What funding strategies can reconcile the need for financial support for public transport with the lack of public sector resources?

These are all big questions that were examined in the research that is reported in this book. By no means have all been answered. However, it was the intention to investigate the issue of the role of local earmarked funding initiatives within a context that reflects a strategic understanding of the causes and magnitude of the transport policy crisis.

Notes

1   In Italy, patronage grew by 170% and in Denmark by 139%. Only in Britain was a decline recorded – of 42%.

2   Noise from traffic meanwhile, is now believed to contribute to stress-related problems such as high blood pressure and minor psychiatric illnesses, and may be an aggravating factor in mental illness (WHO, 1995). The WHO study sees sleep interference as significant. Up to 63% of dwellings are exposed to night-time noise high enough to interfere with sleep.

3   The figures have been converted from US Dollars ($-USD) to € at the rate of €1 to $0.9 (XE, 2001).

4   This calculation assumes an EU-15 GDP of approximately $8000bn during 1997 (IEA, 2001).

5   The concept of managing the demand for transport is one that goes under several names. In the US the term 'Transportation Demand Management' (frequently abbreviated to TDM) is used. In

Australia and the UK (where 'Transportation' can have a different meaning) the variant 'Transport or Travel Demand Management' is used. Among EU policy-makers, 'Mobility Management' is the term most often used. All these terms are largely interchangeable, although they may contain difference in emphasis.

Chapter 2

New finance for public transport

The previous Chapter provided an overview of the transport crisis facing Europe and other developed and developing economies. This Chapter will focus upon one measure that features in virtually all transport policies, which is the development of public transport to provide an alternative to car dependency. The first important point is that the development of public transport *alone* cannot represent an adequate transport policy response. This point was raised at the end of Chapter 1. An integrated portfolio of mobility management transport policy measures is needed to produce any impact upon the current car-oriented regime. This Chapter therefore assumes that the development of public transport would be part of such a holistic approach.

The funding of such public transport developments by local earmarked taxes and charges is, of course, the subject of this book. The purpose of this Chapter is to explore the policy and public finance context in which such mechanisms have emerged.

The role of urban public transport

Public transport services[1] take different forms according to the size, density and history of the area served and can involve a mixture of buses, trams, light or heavy rail and metros. There may also be more innovative systems for niche markets, such as train-taxis, dial-a-ride and people movers. Aircraft and ferries are also public transport, although this book is concentrating upon surface public transport and urban public transport in particular.

For all the reasons explored in Chapter 1, local and national government has come to play an increasingly important role in planning the coverage and nature of public transport systems. Policy goals and the split in responsibilities involved in implementing mechanisms to achieve these vary between countries. National governments often set targets and policy with local or regional government being responsible for delivery and adapting general targets and policy to locally specific situations. The strategic importance of transport policy has now grown so much that transport has even come to feature as an important policy issue at the European Union level.

For the European Commission, transport policy development is governed by two general principles laid down in the Treaty of Europe. These are the subsidiarity principle and the proportionality principle. The first principle stipulates:

Figure 2.1 A Paris Metro heads towards the Arc de Triomphe passing the city's increasingly congested roads

In areas, which do not fall within its exclusive competence, the Community shall take action. (…), only if and in so far as the objectives of the proposed action cannot be sufficiently achieved by the Member States and can therefore, by reason of the scale or effects of the proposed action, be better achieved by the Community.

while the second principle has to be interpreted as follows:

Any action by the Community shall not go beyond what is necessary to achieve the objective of this Treaty.

Within this overall framework a common European policy for the transport sector is being developed.

Urban transport is generally seen as a public service of great importance. At the individual level it is a service that meets the needs of mobility for citizens, while at a societal level it contributes to quality of life and sustainability. The

EU Directorate-General on Transport has positioned public transport as a crucial service for European citizens (see CEC, 1996):

> Needs of citizens are put at the centre of decisions about transport provisions . . .

and

> Ideally, public transport should be accessible, affordable and available to all citizens. Financial and technical considerations may constrain this, but the Commission believes that the goal is important and worth of debate . . .

The specific EU policy goals on public transport are:

◆ to encourage increases in the use of public transport;
◆ to encourage system integration and fulfilment of public service requirements;
◆ to establish incentives for service providers and planning authorities to improve accessibility, efficiency, quality and user friendliness of public transport systems;
◆ to promote financial conditions required for making public transport services more attractive, both for public and private investors;
◆ to ensure minimum requirements in respect of the qualifications of staff, thus guaranteeing high levels of reliability, safety and security;
◆ to safeguard flexibility in relation to specific national regional and local priorities and the particularities of national legal systems.

In the 2001 'White Paper' of the European Commission on 'European Transport Policy for 2010' (CEC, 2001) there is a renewed interest in a Common Transport Policy. This advocates a balanced and cost-efficient approach that introduces, where possible, competition in the transport sector, and that also seeks to exploit all transport opportunities (including public transport) to the maximum extent possible. Although market principles play a crucial role in European transport policy, this is coupled with advocating an integrated portfolio of different measures and flanking policies.

Why intervention?

As noted above, state funding has emerged in response to public transport being used as a policy measure, but it can hardly be said that there is a single policy involved in transport subsidies. Indeed, there are many reasons used to justify subsidy for local public transport infrastructure, fares and service levels. Berechman (1993) notes that the public regulation, financial support and, often, ownership of public transport is usually rationalized on the basis of three major sets of reasons. These are economic grounds (primarily efficiency and equity), political realities (including the power of interest groups), and what is called the social role of public transport. Two of these reasons (economic and social) are rooted in theoretical principles, which interest groups use to advance their case. Basically, the economic justification of market regulation and subsidy of public transport by the public sector stems from two principal theoretical reasons: market failures and income distribution. The following section therefore explores the principles behind public transport subsidy before concentrating on the use of local earmarked taxes and charges as a new revenue source and instrument of transport policy.

Market failures

Market failure-driven subsidies are reflected in two key sets of policies. Firstly there are economic development policies. For example, public transport may be needed to serve a large and congested city efficiently or to 'pump prime' a major new development. A good example of the latter is the construction of London's Docklands Light Rail and the Jubilee tube line to facilitate the massive commercial developments along the old London docks corridor. From the mid 1980s, environmental policy has come to justify subsidy in order to reduce pollution and emissions from the transport sector by transferring demand to public transport. In both cases there is a market failure because the external and long-term costs and benefits are not adequately reflected in market prices.

Market failures occur when competitive markets fail to allocate resources efficiently. This is when the marginal costs faced by individuals in the production or consumption of goods or services differ from social marginal costs or market prices of these outputs. This leads to excessive demand or supply. The most prevalent causes for market failures comprise externalities, public goods, natural monopoly and imperfect information. External effects occur when someone inflicts costs (or benefits) on others, without adequately compensating them for the unmerited effects. Common examples of negative externalities include traffic congestion and noise pollution. In the presence of external effects market prices will (even in highly competitive situations) not be equal to marginal cost. In order to make prices equal to marginal costs, they have to be adjusted by internalizing the external effects. Public transport regulation and subsidy is justified through the role it plays as part of policies to restrain car traffic and so address negative externalities.

The public good argument is applied to many public services. These include the police, army, ambulance and fire services, which are socially and economically necessary but cannot be practically provided on a fee-for-service basis. The provision of infrastructure and transport systems by governments is often explained as a case of a public good (although there are cases of long-distance highways being funded on a toll basis). A result of such a 'public good' is that consumption does not exclude another person from consuming the same level output at zero marginal costs.

The natural monopoly argument results from production under conditions of scale economies. Under these conditions and given the market demand, a good can be produced at least cost if only one firm is engaged in its production. This leads to monopoly power for a firm which may lead to high (above marginal) costs and inefficient prices. These monopolistic rents can be a reason for governments to regulate and improve efficiency.

Imperfect information occurs when costly information leads to high uncertainties and incorrect market decisions. An example of this type of market failure is the perception that the only marginal cost of car use is fuel, leading to overuse. Another example of imperfect information leading to market failure is the often inadequate provision of public transport information (e.g. over complicated, unreliable or non-existent timetables). This may cause under-utilization of the transport service. The provision of public transport by the public sector is sometimes rationalized on the basis of this argument.

Equity and income distribution

Income distribution-driven subsidies are linked to social inclusion policies. For example, providing transport for low income, the elderly, disabled and other 'disadvantaged' groups. Regulatory intervention is also rationalized as a way to affect income distribution. Governments raise taxes to generate revenues for fulfilling their distributional objectives. Some goods and services are considered essential for the basic welfare of individuals; a minimum level of supply of these goods to all inhabitants of an area is then an objective of governments. The consumption of these goods is perceived as a basic civil right (or need) of all individuals, hence a 'merit good'. Transport is one of these goods that should be available to all people. Transport typically accounts for 19% of household expenditures and is needed to get to jobs, education, health care, and other basic services (Gómez-Ibáñez, 1999). As lower income groups have less access to cars, they are dependent upon public transport for their mobility needs. As car use has eroded the financial viability of public transport services, subsidies have been justified to protect low income and other vulnerable groups.

These theoretical principles are not just of academic and macro-economic interest. There are major practical implications for the design of the subsidy funding mechanisms. If the core rationale for public transport subsidy is rooted in policies for income distribution and social inclusion, then the funding sources are likely to reflect these goals. There would be no reason to consider any of the market failure issues. Thus a socially-driven subsidy policy would place emphasis on reducing access barriers to public transport (fares, service levels and physical access) and funding mechanisms might also seek to achieve income redistribution (sources would be likely to be diverse and progressive, perhaps including taxes or charges upon income, consumer goods or services).

By way of contrast, if the rationale were economic, this could well be reflected in the design of a mechanism to capture income from those who received the external benefits of public transport in order to fund its costs. This may include taxing or charging those who benefit from the enhanced level of public transport, such as employers or property developers.

Justifying subsidy

Subsidies are often criticized as being inefficient and unfair (Litman, 2002). Critics argue (*inter alia*) that public subsidies encourage inefficient public transport service, increasing costs and that they are unfair to people who pay for services they do not use. However, subsidies can be justified on several grounds which are related to the reasons discussed at the beginning of this chapter, including generally equity, the economies of scale argument, and considerations of second best solutions. These will be discussed in more detail below.

Equity

Equity and social inclusion constitute a major justification for subsidizing public transport. Subsidy allows the redistribution of income to transport disadvantaged groups and economically backward regions, by transferring real income in the form of public transport services rather than cash. Reducing public transport fares by subsidies helps people who do not own a car, providing a basic level of mobility and access to employment and services. An equity justification for subsidy frequently focuses upon key groups such as the poor, unemployed, the disabled, the young and the elderly. It also applies to certain areas, including places hit by high unemployment and rural areas, although in the latter case political influence and a wider rural agenda can result in a high level of subsidy beyond economic rationale. For instance, rail transport service in Germany is even provided in regions where patronage is less than 1000 people per day (Rothengatter, 2001). Critics of the equity justification note that it depends on the extent to which these groups use public transport, because subsidies are primarily going to the providers and not the end users. And when users benefit from lower fares, it is often not the poor who benefit. For example, in the largest and densest US metropolitan areas, the average household income of urban public transport users is similar to the average household income of all metropolitan residents because transit patronage is dominated by commuters to the central business district, many of whom are highly paid (Gómez-Ibáñez, 1999). It may be argued that the intended redistribution could be accomplished in a less distorting way. At the EU level, the role of public transport is extended to include not simply a narrow equity and social inclusion aspect, but is seen as offering effective and novel contributions to social and economic objectives of the EU, as formulated in the European Commission's 'Citizens Network' initiative (CEC, 1996).

Second-best solution

Political realities can affect subsidy policy in a number of ways, in particular, when economic pricing cannot be achieved in practice. Many current transport problems (excessive congestion, pollution emissions) result from market distortions and failures that encourage automobile use (as discussed earlier and also in Litman, 2002). In a more economically efficient transport system motorists would face higher charges (especially for peak hour driving, which is priced below marginal social costs), and so would have incentives to limit their vehicle use and use alternative modes. If it is impractical for road use to be fairly and efficiently priced, there is an argument for subsidizing its substitute. Subsidies are therefore a 'second best' but politically necessary compensatory solution. When some people switch from the car to public transport, this reduces pollution, congestion and noise, which benefits other people who are not public transport users. External costs are lowered and therefore public transport should be subsidized. This is a crucial argument for subsidy as a transport policy measure. If it is politically impossible to price another public good (road access by cars) at the true level, then subsidizing the alternative (public transport) is an alternative way to achieve the same effect.

An argument has been made that current public transport subsidies are excessive and economic efficiency would be served better by decreased rather than increased subsidies (Proost et al., 1999).

Decreased public transport subsidies and higher fares would need to be accompanied by higher pricing of road access, but therein lies the difficulty of this argument. In practice it is not politically possible to price road access realistically and so public transport subsidies are used to compensate for road access subsidies. The issue of internalizing transport externalities is the core of the European Commission's Green Paper, 'Towards Fair and Efficient Pricing' (CEC, 1995) and its White Paper, 'Fair Payment for Infrastructure Use' (CEC, 1998). These advocate that infrastructure charges should ideally and normally be based on marginal social costs at the level of actual use. These marginal social costs should also incorporate all external costs, such as congestion, pollution, and accidents.

Economies of scale

There is a market-based argument that public transport is a 'natural monopoly'. Subsidies may be justified for goods and services that experience economies of scale, implying that marginal costs are lower than average costs. Economies of scale may not only lead to monopoly, but also make it difficult for private operators to stay in business in the public transport sector. Public transport service tends to have such a cost structure, since once a service is established, the marginal cost of accommodating an additional rider is rather low, and as ridership increases, the scope and frequency of services can increase (Litman, 2002). However, econometric studies of bus and rail operators generally show little evidence of scale economies (Gómez-Ibáñez, 1999). Fixed costs are large, especially for rail systems, and variable costs are relatively small. Public transport companies usually operate at less than capacity (they could carry more riders with little increase in costs); and marginal costs for an extra passenger are consequently low. Therefore, if price is set equal to the marginal cost (this is the economic rule in price setting as being the most efficient allocation of resources and maximizing social welfare), public transport operators will suffer a loss, because the marginal cost is less than the average cost. Clearly, funds must be raised somewhere to keep these operators in business, and this financial support is often provided in the form of a subsidy if public transport is deemed to be in the public interest.

As well as the issue of the justification for public transport subsidy, there is the issue of the optimal amount of subsidy. Several economists have made theoretical derivations of optimal public transport fares. Attempts to devise operational pricing rules for urban transport are, for example, reviewed by Nash (1988), while subsidy policy is discussed by Else (1992). Their recommendations are difficult to implement, as it is basically a political decision as to what amount of subsidy is given. In practice, there is a wide variation in subsidy policies depending on a host of local, political and operational circumstances.

What and how to subsidize

Apart from why public transport should be subsidized and at what level, another question is exactly *what* should be subsidized. One view is that government subsidies for *infrastructure provision* (e.g. rail) are defensible on efficiency grounds: a considerable share of the losses with optimal marginal cost pricing may result from the large fixed cost of infrastructure. However, the *operation of the services* may in many circumstances involve constant or even decreasing returns to scale, in which case one of the major economic motivations for subsidies would vanish.[2] Deregulation and privatization policies may then involve the auctioning of the rights to operate a service to private companies (see Small and Gómez-Ibáñez, 1999), which illustrates the point that increasing state involvement and financing of public transport does not of itself require state ownership. A number of regulatory/ownership arrangements exist across the EU and worldwide. In all cases it has involved the local or national state providing finance to develop and maintain public transport to the desired level, but this can also involve a whole variety of public and private ownership structures.

In general, one can distinguish between provider-side subsidies and user-side subsidies. Provider-side subsidies involve financial support for the provider of the public transport service, for example covering the deficits of a company. But it is also possible that governments attach strings to the support of firms. It is, for instance, possible that governments decide on the tariff structure or demand a certain quality of service (e.g. transport in particular areas). A subsidy can also depend on the performance (e.g. number of kilometres or number of travellers) of the public transport operator. An example is the UK minimum service and reliability specifications for subsidies to franchised rail operators. Subsidies can also be offered in a more hidden way. Tax exemptions may be provided for public transport (e.g. from vehicle and fuel taxes or reduced ecological taxes).

User-side subsidies are less common. Subsidies are not given to the operator, it is the traveller who gets the benefits directly. The operator gets the subsidy only when the traveller makes use of its services, so it may be tempted to meet the client's wishes. Another advantage is the flexibility for the government to structure the subsidy. It is possible to address the subsidy to particular groups in society, such as the disabled, the elderly and the poor. Public transport passes or travel allowances for such groups are commonplace. Again there may be tax allowances rather than direct subsidy. For example public transport costs may be an allowable income tax deduction for commuting.

Public transport subsidy in practice

In practice, the level of subsidy to public transport varies significantly between countries. In Europe, the level of government subsidy for buses varies between 30% and 70% of total operating costs (Figure 2.2). Austria, Belgium, and Italy are the top three ranked countries in terms of reported urban subsidies/ grants (in the range 60–70%). The United Kingdom has the lowest average subsidy rate at just over 30%.

In the United States, subsidies fund up to 75% of transit expenses (see Table 2.1).

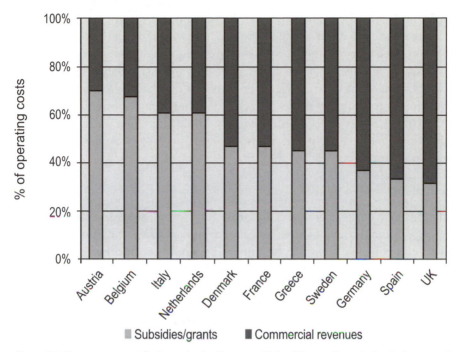

Figure 2.2 Revenue support for buses in the European Union (*Source:* Commission for Integrated Transport, 2001)

This variation in subsidy level is due to a variety of reasons. Although cost structures and efficiency are important issues, relative fare levels explain a large amount of this subsidy difference. As noted earlier, market failure is a major justification for public transport subsidy. Many socially necessary public services are 'public goods' that collect no revenues and are state-funded and paid for on a collective basis. This has historically included most roads and some transport services. Public transport does collect revenue (through passenger fares) but this is not usually sufficient to cover costs, particularly when fares are set low in consequence of a social inclusion view of its

Table 2.1 Government subsidies to public transport and roads in the US, 1970–1996 (all government levels combined, in millions of inflation-adjusted, constant 1996 dollars)

	1970	1980	1990	1996
Subsidy to public transport	2691	15,311	17,718	20,473
Percent of total public transport costs covered by passenger fares	74	26	31	28

Note
Subsidy includes both operating and capital subsidies for all forms of urban public transport (94% of total subsidy) and for Amtrak (6% of subsidy). Subsidies to intercity bus services in the US are negligible.
Source: Pucher, 1999

provision. If a major rationale for public transport is to provide mobility for lower income and disadvantaged groups, fares need to be low enough for them to use the services regularly. Low fares can also be necessary for public transport to be effective as an instrument of transport policy. To attract car users, fares need, at most, to be comparable to the perceived cost of driving. This equates to the cost of fuel, which varies between nations. In particular, in the United States and Canada, the very low cost of fuel means that to be competitive in any way, public transport fares need to be lower than in the higher fuel cost regime of Europe.

Revenue and capital subsidy

Some bus systems are commercially provided, but in others fare revenues can cover operational costs but not the cost of capital replacement. Subsidies (often called 'grants') therefore arise for capital expenditure. In other cases where, for example, there is a policy for low fares, there are revenue subsidies as well. In many countries there is a different arrangement for capital and revenue subsidy, with local authorities often expected to finance revenue subsidies. In the United States, most capital funding for urban transport comes from the Federal Government, while fares and local, regional and/or State sources cover operational costs. One blurred area regards the purchase of new vehicles. In the UK/Europe, replacing vehicles is considered to be an operational expense, whereas in the US it is covered by capital revenues. In Britain, private bus operators cover the operational costs through fare revenues. The majority of bus services are commercial and provided like any other private sector service. Subsidy is concentrated upon a minority of loss-making 'socially necessary services' and discounted fares for the elderly and disabled. This is the responsibility of local government, which is also responsible for providing on-street infrastructure and sometimes bus stations (depots and most bus stations are part of the private operator's commercial costs).

The relative balance between capital and revenue costs is different for rail than for bus systems (see Table 2.2). For bus, the vast bulk of costs are variable, whereas for rail the majority of costs are fixed.

Table 2.2 Typical distribution of bus and rail costs

	Bus		Rail	
Variable costs	Crew, fuel, tyres, oil	55%	Crew and fuel	25%
Semi variable costs	Vehicle maintenance, depreciation	25%	Vehicle maintenance, depreciation	20%
Fixed costs	Garages, overheads	20%	Terminals (15%), track and signalling (25%), administration and general (15%)	60%

Source: TRRL, 1980

Many countries have special arrangements for capital grants. These typically fund the construction of a new rail line, installing new signalling, the construction of a busway or bus priority infrastructure. Such grants are often provided by central rather than local government.

In some countries, the split between capital and revenue support is less crucial. This can apply where all subsidy is integrated into one tier of government. This happens in France and also in Canada where both public transport operations and infrastructure are typically the responsibility of the Province, or in larger cities, e.g. Vancouver, Montreal, Ottawa and Toronto, regional agencies. Where there are franchise arrangements, as in some Dutch cities and for rail in the UK, capital and revenue may be integrated, although with some exceptions made, for example for the very large rail infrastructure projects in the UK which span several franchise periods. There is also a variety of systems to raise large sums for capital expenditure, which is then repaid via a revenue stream, and this also blurs the distinction between capital and revenue expenditure. In the United States this commonly takes the form of the issuing of public bonds which are repaid from a revenue source. Very frequently the revenue source is a local hypothecated tax or charge. In contrast to this public sector-controlled process are the Private Public Partnerships in the UK, such as that agreed in 2002 to refurbish and upgrade the London Underground. Under this a private consortium is given a lease upon infrastructure, has a contract to improve and maintain it over that period, and is paid an annual charge for doing so. The private consortium therefore raises the capital needed and manages the whole process.

Sources of subsidy

As noted above, the money for subsidy can come from different sources depending very much on the institutional organization. In particular, it appears that public transport is funded differently when comparing the United States with Europe. Subsidies in the United States come from Federal, State and local funds (see Table 2.3). This overview shows that transit funding increased over time to reach a total of almost $23 billion in 1994. More important is the role of local taxes earmarked for the sole purpose of supporting transit. These are the second most important source of revenue after fares. Dedicated funding for operations and capital expenditures (which includes dedicated taxes and other dedicated funds at the State, local, and agency-jurisdiction levels) has become the fastest growing component of funding. This indicates a significant shift of funding responsibility directly to the communities that benefit from transit. These local option taxes are usually authorized by State legislation and often require a referendum for adoption. Local governments, together with public transport companies, need to communicate the objective of the tax in order to win the passage of the referendum. By contrast, State and local funding for public transport operations from general revenue sources (which is not voter approved) has actually dropped in nominal dollars.

As opposed to operations, capital investments have largely been funded by the Federal government. However, as with the operating funding environment, a fairly notable shift from reliance on Federal funds to dedicated taxes can be identified. More recent experience (1994 to 1998) has shown continued decreases in Federal funding (and Federal operating funding in particular) (TCRP, 1998). As a result, many public transport companies have been compelled to adjust service levels and modify funding strategies by increasing State and local shares and looking to non-traditional revenue sources.

This approach in the United States of local organization and funding is a less known phenomenon in Europe. Some countries do have a high level of decentralization, but these rely most often on national governmental financial support (such as Italy). In small highly urbanized states such as Belgium and the Netherlands, public transport is usually organized and funded nationally (Farrell, 1999a). In southern Europe public transport is a local responsibility, but funding largely comes from the central government. Political decentralization is also apparent in Scandinavia and the Alpine countries (Switzerland and Austria), where funding is left to regional authorities (e.g. cantons in Switzerland). The use of dedicated taxes to fund public transport is a rare phenomenon in Europe.

Traditionally, additional funding has been raised in a number of ways. Firstly, public transport operators have sought to maximize revenue from other complementary activities such as hiring vehicles for private charters, selling advertising space on vehicles or facilities such as stations or depots, or through renting or selling property.

Nakagawa and Matsunaka (1997) note that ideally a combination of methods should be used to fund public transport systems, such that:

Table 2.3 US operating and capital revenue trends (in billions of nominal dollars)

Source	1989	1990	1991	1992	1993	1994
Fares	5.11	5.51	5.60	6.24	6.53	6.47
Other revenues	0.72	0.81	0.75	0.60	0.54	0.97
Federal	3.10	3.45	3.39	3.44	3.30	3.37
State	2.24	2.44	3.71	4.31	2.81	2.53
Local	2.50	2.79	4.63	3.99	2.94	2.73
Dedicated	4.08	4.57	2.99	3.15	6.37	6.87
Total	17.75	19.57	21.07	21.73	22.49	22.94

Source: TRB, 1998

- investment should match need;
- projects should be efficiently realized; and
- the burden should be fairly and equitably distributed.

Drawing on this framework, Nakagawa and Matsunaka (1997) go on to suggest that:

- The size of revenues from *user fees* depend on customer choice so pricing must be competitive. By the same token, unnecessary investment of resources are discouraged. But if a transport system depends entirely on this revenue, profitability becomes the sole basis of decision-making and social benefits and costs are ignored. External diseconomies lead to excessive investment, and external economies result in insufficient investment. Such investment is likely to be concentrated in large cities where profitability is higher than in less densely populated areas.

- The use of *public funds* is the preferred method of financing public transport when externalities are present or when fundamental social rights need to be guaranteed. Use of public funds is a political decision and can satisfy the principle of fairness, unless the decision-making process is inadequate and arbitrary. Generally speaking, work that is state-funded involves little incentive for profit making and for this reason tends to be relatively inefficient.

- Finally, the use of *long term debt* means that the distribution of burden between present and future generations may achieve a degree of fairness, although future generations cannot be consulted. The level of investment is based on uncertain cost and benefit estimates and the accuracy of these estimates affects the quality of decision-making. Excessive investment will be undertaken if future benefits are overestimated; inadequate investment will result from underestimation. Uncertainty makes effective decision-making a highly difficult task.

New sources of funding

Today, public transport operators receive significant subsidies. This is particularly so for public transport systems with heavy infrastructure costs, such as rail, metro and light rail systems. Even for public transport systems with low infrastructure need, such as bus, there is the labour-intensive nature of the industry and the increasing maintenance needs of the older systems. There have also been trends that increase costs. The pattern of demand has changed with the suburbanization of jobs and residences and development of car-oriented land-use patterns. This makes it increasingly difficult and costly to provide good quality public transport. All these have combined to burden many agencies' cost and revenue structures (TCRP, 1998). This has led to a general apprehension about the growing gap between operating expenses and revenues.

The need for additional funding is in contrast with the recent trend of reduced government financial support for public services (especially in Europe and North America). The use of general tax revenues for public transport support and development is becoming more constrained and uncertain. Efficiency savings, often involving privatization, have been one response although in some cases (e.g. British Rail) this has ended up substantially *increasing* the level of state subsidy. An approach, which can be combined with efficiency savings, is for public authorities (often together with public transport operators) to develop alternative sources of funding. Innovative funding techniques may include the developing of non-farebox revenue, adopting private sector methods (e.g. turnkey development), new fare structures, value capture strategies, use of property rights, leasing techniques, and dedicated (local) taxation sources.

Figure 2.3 New Metro Station in Los Angeles – a classic case of difficult and expensive public transport operating territory

Public transport is thus facing a paradoxical situation of being required to meet an increasing number of economic, environmental and social objectives on an ever decreasing level of 'conventional' public money. To resolve this situation, either production costs need to be cut (through efficiency gains of one sort or another), and/or new sources of finance need to be found.

Using local earmarked taxes to finance local public transport

The idea of collecting a tax or charge for a dedicated purpose is not new. For example, this was the principle of the UK's original *Road Fund Licence* introduced early in the last century. Motorists were required to pay this to help fund the upgrading of roads to standards needed for motor vehicles. The earmarking of the funds to road improvements eventually ceased and the fund (later renamed Vehicle Excise Duty) became a general source of taxation revenue. National taxation dedicated to transport purposes is rare in modern economies, but it is at the local level that this funding mechanism is being increasingly integrated into transport policy.

There is a variety of terms to describe what this book calls *Local Earmarked Taxes* (LETs). The idea of the revenue being dedicated to a particular purpose is sometimes referred to as 'earmarked ', 'ringfenced' or 'hypothecated', the last being the technical term used by finance ministries. Whatever it is called, the mechanism by which the dedicated money is raised may be a tax, a charge, or a levy, depending on its legal status. Whatever they are called, the use of LETs in this book involves a wide number of local taxes, charges and levies, some or all of the revenue from which is directly earmarked to fund public transport. There are some LETs that only fund road improvements, which will not be covered in this book, as the focus is upon their use in demand management transport planning. Although, in recent years, the use of LETs for local transport demand management has attracted growing attention, (e.g road pricing) in general the more established LETs were developed simply as a source of income to support public transport services or to fund their expansion. This is linked to two trends in local public transport finance and provision.

The first, as has been noted already, relates to difficulties with the traditional forms of financing public transport investment, i.e. grants to municipalities from national government. The second factor is that there has been a trend in a number of EU Member States, and elsewhere, to devolve the responsibility for local and regional public transport away from national government. This has led to the desire to devolve funding mechanisms too, which in some cases has involved the development of LETs. This could be viewed as an example of real national subsidiarity in action. It is *real* subsidiarity because not only are responsibilities devolved to a lower level of governance, but also powers and finance (the latter being notoriously absent in some so-called examples of devolution).

In countries with a federal system, the local state or region usually has some form of local taxation within its control. In such cases, the introduction of LETs need not involve any new funding legislation, but can be accommodated within existing structures. As is noted throughout this book, the United States is a prime example of this. In response to a reduction in Federal support for public transport, individual states and cities used their powers to raise a whole variety of local taxes in order to support and develop public transport systems. In other cases, special charging powers have been made available to introduce LETs, such as the *Versement Transport* tax in France and the *Congestion Charge* in the UK.

The case for and against dedicated taxes and charges

The dedication of revenue streams is something that finance ministries have long disliked and there is considerable institutional opposition to it. Deran (1965), in a classic text on the issue of earmarking revenue streams, summarized the criticisms and justifications for earmarking. The criticisms were that:

+ earmarking hampers effective budgetary control;

Figure 2.4 Vehicles entering the London Congestion Charging Zone. Note the road markings and gantry with number plate recognition cameras

- earmarking leads to a misallocation of funds, giving excess revenues to some functions while others are under supported;
- earmarking impacts inflexibility to the revenue structure, with the result that legislatures are hard put to make suitable adjustments when conditions change;
- earmarking provisions often remain in force long after the need for which they have been established has vanished;
- by removing a portion of fiscal action from periodic review and control, earmarking infringes on the policy-making powers of state executives and legislatures.

However, more positively, he also found clear justifications for earmarking, in that:

- earmarking can apply the benefit theory of taxation;
- earmarking assures a minimum level of expenditure for desirable governmental functions, avoiding the need for wasteful repeated pressures on the legislature;
- earmarking, by assuring continuity for specific projects, can reduce the cost of these projects through lowered bond interest rates and advantages of long term planning;
- earmarking can help overcome resistance to new taxes or increased rates.
- earmarking is a proven way to influence citizens' acceptance of measures.

The justifications clearly strike a chord in relation to the funding needs of public transport, particularly when capital projects are involved. In a more recent commentary on the public finance principles of earmarking, Teja and Bracewell-Milnes (1991) conclude that:

> the traditional objections to earmarking are weak and invalid because they assume a Utopian system of public finance and democratic decision-taking that bears little or no relation to reality. Earmarking is an exercise in the second-best or least bad; in an imperfect world, it can provide better decisions and do less damage to the creation of wealth than conventional pooled financing of government expenditures.

They continue 'earmarking creates wealth in two separate ways: by improving the allocation of resources and by giving scope to the voluntary principle. In each of these ways wealth is created through the replacement of compulsion by choice'.

Overall it appears that public transport is one area of public expenditure where revenue earmarking is particularly appropriate, although earmarking should not be taken to an extent that it causes problems of inflexibility in public finance.

Dedicated taxes and charges and principles of taxation

Once the concept of earmarked taxes and charges to subsidize public transport has been accepted, the next issue concerns the source of this revenue. This concerns the principles behind how this money is raised, and these principles need to be considered whether the mechanism is labelled a 'tax' a 'charge' or a 'levy'.

> The purpose of taxation is to raise revenue to finance government spending . . . The design of the tax system affects the proportion of total revenues borne by different groups or activities, and it affects economic activity, as individuals and firms react to the taxes they face. Because of this, taxes can also have other purposes: to influence the distribution of income and to influence economic and social behaviour in various ways. (Commission on Taxation and Citizenship, 2000)

This quote shows that taxation has served four major purposes:

(a)   to raise government revenue;
(b)   to pay for specific collective goods and services;
(c)   as an instrument of economic policy;
(d)   as an instrument of other policy areas.

Road transport taxation was initially introduced for (b) (and this is behind dedicated transport taxes in many countries). It reflects the 'public good' aspect of transport infrastructure, but there is no reason why general taxation cannot be the source of income, and so it soon merged into (a), which is the longest-established rationale for taxation.

The third rationale (c) emerged in the wake of Keynesian macro-economic theory after the Second World War and, informed by various subsequent economic philosophies, has been with us (in some form or another) ever since. The fourth rationale (d), that the design and implementation of taxation measures should serve other policy aims is a recent and only tentatively established purpose of taxation, which includes the use of fiscal instruments for environmental policy. This is linked to the 'market failure' justification for subsidy. LETs, may have an element of all of these purposes.

Whatever the purpose of revenue raising, there are long-established criteria in the design of a measure. In general, to be accepted, taxes must be perceived by the public as being:[3]

- legitimate – both in purpose and operation;
- progressive – the tax system should take a higher proportion of income from those on higher incomes and wealth than poorer people;
- economically efficient – it should encourage work enterprise, saving and economic efficiency;
- discourage social harm – incentivize reducing socially damaging behaviour;
- be equitable – i.e. treat people in similar circumstances similarly;
- tax individuals separately;
- have a broad base;
- administratively efficient – cost effective to collect and enforce; and
- raise sufficient revenue to fund the desired level of expenditure.

Approaches to implementing LETs

Local earmarked taxes and charges have emerged over a number of years and in different specific situations. It is therefore not surprising that the design of these various LETs measures has placed different emphasis on the principles and purposes of taxation considered above. There appear to be three main groupings, depending upon who pays and the purpose for which the revenue is being gathered. These are:

- Beneficiary Pays
- Polluter Pays
- Spreading the Burden

These categories tend to map onto contrasting rationales for public transport subsidy. *The Beneficiary Pays* group appears to reflect a view that the LET is to pay for collective 'public goods'; *the Polluter Pays* LETs can be (though not always) used as an instrument of environmental/transport policy in order to discourage social harm; *Spreading the Burden* tends to involve LETs with a socially-driven subsidy policy. This section will introduce the principles behind these three groups of LETs measures. Chapters 3, 4 and 5 of this book will then provide detailed examples of the practical design and implementation of these three groups of measures. Some LETs display characteristics of more than one of these groups. However this categorization, in that it provides links to both the theoretical basis of subsidy and practicalities of policy/instrument design, is seen as a useful structure.

Beneficiary pays (public good and 'economic' rationale)

As noted previously, one of the oldest rationales for taxation is for public goods that cannot be provided on a market exchange basis. Defence, police, ambulance and fire services are obvious examples as are most roads (although it is possible for motorways to be provided on a market basis using tolls). This includes certain aspects of public transport, such as the economic benefits to towns and cities and labour force benefits to employers.

Because public transport is seen as providing a collective public good benefit, LETs are used to charge people and organizations for these collective benefits. This could involve a local charge to the area where public transport investment takes place or to employers located in that area. Chapter 3 contains a number of examples of LETs that have developed from this approach. These include LETs on employment, on property, on land values, and on developers. Of these, the last is most common, with payments from developers being required to compensate for the transport impacts generated.

An example of a recent approach using the beneficiary pays principle is a study commissioned by the Royal Institute of Chartered Surveyors looking at how planned improvements to London's transport infrastructure could be financed (Whelan, 2003). The basis for this is 'that Government is likely to require some contribution towards certain transport schemes' and therefore the study explores 'innovative funding methods' that could help fund developments. The study particularly focuses on property taxes, because the property value impacts of the public transport schemes were estimated to be significant. This includes a consideration of a business rate levy, tax incremental financing, Business Improvement Districts, land value taxation, and greenfield development tax. The potential yield of the LETs considered is estimated to be between £10m–£450m (€14m– €640m) per annum.

Polluter pays (market failure and environmental rationale)

The 'polluter pays principle' (PPP) is a more recent, and less well established, approach to both national and local taxation. Not only is it a different perspective but it is advocated by a different set of actors. Transport and environment ministries have seen LETs as a tool for transport demand management. As a result, a new group of LETs has emerged from a transport/ environmental policy perspective, and these relate to a different set of public finance and economic theory issues. This centres upon the aim of 'discouraging social harm' by using the tax system to address environmental externalities.

The issue of external costs and benefits is long-established in economics. Transport activities are a prime example, with the costs not entirely borne by individuals involved in undertaking their travel. The costs associated with air and noise pollution for instance are not taken into account in deciding how many journeys to make, because either the travellers are unwilling to do so or they are unaware of them. These external costs of transport cover a number of factors, including not only air pollution and climate change gases, but also accident and congestion costs (as was discussed in Chapter 1). Because of these externalities, the transport market is not operating in an economically efficient manner. Economic theory suggests regulation to rectify this position, or measures to internalize the external costs. The most efficient solution is the imposition of a Pigouvian tax (i.e. a tax that is levied on each unit of a polluter's output in an amount just equal to the marginal damage it inflicts *at the efficient level of output*). Imposing such an ideal tax is still not politically realistic in transport practice. But, it is clear that governmental intervention is justified, in some way, to compensate for the effect of transport's external costs.

There are several possibilities for authorities to intervene in the transport market in order to create greater efficiency in its functioning. A wide variety of policy instruments can be used to address the environmental costs of transport activities, including emissions fees and tradable permits. Table 2.4 provides a list of such measures.

In practice, taxes and charges on fuels and vehicles are the most commonly used measures. Historically, such taxation has existed for many years as a reliable source of general government revenue. Many may have started, as already noted above, as dedicated taxes to fund road building, but then became incorporated into general government finance. Now, with the rise of environmental concerns, transport taxation has come to be justified in terms of environmental policy. The basic tenet is that the price of a good or service should fully reflect the total cost of production and consumption. The use of

Table 2.4 Taxonomy of policy instruments to control the environmental impacts of motor vehicles

	Market-based incentives		Command and control regulations	
	direct	indirect	direct	indirect
Vehicle	◆ Emission fees	◆ Tradable permits ◆ Differential vehicle taxation ◆ Tax allowances for new vehicles	◆ Emission standards	◆ Compulsory inspection and maintenance of emissions control systems ◆ Mandatory use of low polluting vehicles
Fuel		◆ Differential fuel taxation ◆ High fuel taxes	◆ Fuel composition ◆ Phasing out of high polluting fuels	◆ Fuel economy standards ◆ Speed limits
Traffic		◆ Congestion charges ◆ Parking charges ◆ Subsidies for less polluting modes	◆ Physical restraint of traffic ◆ Designated routes	◆ Restraints on vehicle use ◆ Bus lanes and other priorities

Source: Button and Rietveld, 1993

market-oriented instruments such as levies and tradable emission rights within environmental policy is the clearest and most direct way of interpreting the 'polluter pays' principle. By the use of one or a combination of the measures from Table 2.4, negative effects of a product or production process are internalized in the costs of products and services.

This is the concept behind the European Commission's green paper on *Fair and Efficient Pricing for Transport* (CEC, 1995) and a variety of national measures that have seen taxation varied according to environmental impact, for example, favouring lead-free petrol, more fuel-efficient vehicles, and encouraging commuting by 'greener' travel modes.

The CEC green paper notes a variety of substantial external costs not directly born by users (congestion, pollution, health etc.). Rather than the traditional approach of addressing these issues by regulation, the paper explores ways of making transport-pricing systems fairer and more efficient. This approach is supported in a Fabian Society text on taxation policy in Britain (Commission on Taxation and Citizenship, 2000), which notes a number of advantages of using tax over regulation as a way of influencing behaviour.

> In principle there are a number of advantages of taxes over other instruments. In particular, a tax is likely to be a more efficient method than a legal regulation. Unlike a legal regulation, which forces all firms to behave in exactly the same way, a tax allows them to choose their own response to the measure according to the costs of doing so. Firms which find it expensive to reduce the damaging activity will prefer to pay the tax, while those for whom reducing pollution is cheap will cut their damage further. This means that the goal is reached at the lowest total cost; that is, in the most efficient way. The uniform behaviour change of a legal regulation will generally have higher costs. A further advantage of taxes is that they encourage consumers continually to reduce their environmental impacts. Since every additional reduction will reduce the tax bill, there is always an incentive to cut further. Uniform standards by contrast provide no incentive to reduce damage beyond the standard set. This is a very important advantage, since one of the main motors of environmental improvement is innovation: improvements in technology and organisation which increase efficiency. Innovation almost always requires investment. This is likely to be encouraged where there are clear and ongoing financial benefits. (Commission on Taxation and Citizenship, 2000)

A further important principle of the CEC green paper is the need for more differentiation, as the external costs vary across space, time and modes.

> The proposed pricing strategy . . . should fully take account of local circumstances. This is important for reasons of efficiency and equity.

In this respect local LETs seem to have the potential to address geographically specific issues (which is difficult if applied at the national level) and many could be targeted by time of day and by mode. Others, such as congestion charges, have very specific targeting built into their design.

The CEC green paper also identified *transparency* as an important issue, noting that:

> accounts should be published identifying the relation between charges and costs. The principal aim of such a policy would not be to raise tax revenues, but to use price signals to curb congestion, accidents and pollution. If this policy were successful, revenues from charges would fall. (CEC, 1995)

The CEC green paper further emphasized the economic gains that *Fair and Efficient Pricing* would produce if it successfully reduced the cost of air pollution, accidents and congestion. It was also necessary if the internal market in transport is to be achieved across the EU. However, it was also recognized that, by their nature, many measures to implement *Fair and Efficient Pricing* would need to be done by Member States.

The concept that a taxation system should be used to take full account of the external environmental costs of economic and consumption activities reaches its logical consequence in the principle of *Ecological Taxation Reform (ETR)*. The concept of ETR was developed by German and Dutch authors in the late 1980s (Von Weizsäcker, Bleijenberg and Sips) and is defined by Whitelegg (1992) as:

> based on the principle that taxes should fall most heavily on those activities and materials that produce pollution and/or environmental damage. Such taxes would replace taxes on labour and capital . . . The total taxation burden in Europe would remain constant. ERT is not an additional tax; it is a replacement tax.

The core of thinking behind ETR dismisses the concept that environmental concerns can simply be added on to existing fiscal structures, but that these structures themselves need to be subject to ecological reform. Our taxation and fiscal systems contain many elements that produce adverse environmental impacts; simply adding on a few ecotaxes will fail to address this structural effect.

On many different levels the current taxation system in the EU is contrary to such ecological pricing. It is usually cheaper to develop greenfield sites than it is to re-use former industrial, military or commercial sites. It is cheaper (or perceived to be) to use the car for routine journeys than it is to use public transport. And it is cheaper to incinerate and landfill waste than it is to develop an effective materials economy where the use of virgin raw materials is seen as a last resort. The current taxation system steers the economy in a very clear direction that is largely non-sustainable. Instead of adding ecotaxes to the existing taxation regime, under ETR, the whole system of taxation would be changed to taxation according to environmental impact.

Ecological Taxation Reform has some clearly defined characteristics:

◆   its objectives are to steer the whole economy in the direction of greater environmental and ecological efficiency;
◆   there is no increase in overall tax take;
◆   ecological taxes replace taxes on labour and VAT;
◆   new taxes are imposed on materials, waste, pollution, water and energy; and
◆   taxes are adjusted to favour re-use of land and discourage use of greenfield sites.

It must be noted that Ecological Taxation Reform is more comprehensive and radical than the EC's *Fair and Efficient Pricing* proposals, although these could be viewed as an example of Ecological Tax Reform in the transport sector.   However, the whole concept of Ecological Taxation Reform is a change in the basis of taxation, rather than superimposing some new criteria on selective parts of the existing tax system.   An important consideration is that an isolated charge or tax, however well designed, cannot successfully influence travel behaviour if the rest of the fiscal and regulatory system is operating contrary to it. For example changes to transport prices may be entirely counterbalanced by existing pricing making car dependent city-edge developments cheaper than low car transport dependency city centre sites.

The European Parliament supported the principle of Ecological Taxation Reform in its discussions and report *Fiscal and Economic Instruments of Environment Policy* in 1991. The EC's 5th Environmental Action programme supports the principle of internalizing external costs ('getting the prices right') and, as noted here, has pursued 'fair and efficient pricing' in transport. The UK's sustainable development strategy also endorses this principle. However, practical progress with Ecological Taxation Reform in the EU has been very slow. The existing taxation system in most countries is characterized by substantial inertia and proposals for change are resisted. Elements of Ecological Taxation Reform are appearing as additional tax measures or as piecemeal reforms to environmental hotspots. Denmark and Sweden have moved the furthest in this direction in linking new environmental taxes to reductions in the level of income tax. The UK has a landfill tax, had a fuel price escalator, and both annual car excise duty and company car taxation have been reformed to vary by CO_2 emissions, though the UK remains resistant to general carbon taxation. Scandinavian countries have elements of carbon taxation and Germany has tax differentials for cars with catalytic converters that have resulted in the rapid reduction on non-catalyst fitted cars. The latter is regulated at federal level but administered at the regional level.

Von Weizsäcker and Jesinghaus (1992) argue that Ecological Taxation Reform presents a complete and holistic approach to the most pressing needs of adjustment and ecological modernization in the search for policy measures that can deliver high quality living environments, a sound economy, and eliminate pressing environmental problems, especially climate change. Indeed they argue that Ecological Taxation Reform has the potential to supply jobs at the level of local economies that conventional macro-economic policies cannot supply. More importantly in our consideration of LETs, Ecological Taxation Reform has the potential to deliver a consistent and integrated fiscal policy that can be fine-tuned to bring land-use systems, individual choice and the supply of transport infrastructure into line with one another.

The dangers of contradictory, self-defeating and self-cancelling policies are very real indeed and great care is required in the design of new measures if these are to be grafted onto an existing system that is not intended to provide an environmentally-beneficial 'steer'. The danger is that if LETs approaches are superimposed on a largely non-sustainable taxation system great inefficiencies can be incurred through different taxation regimes cancelling each other out or pulling against each other. In addition, confusing signals are sent to the principal actors and users at the points where they make their decisions, for example on whether or not to purchase a first, second or third car and where to live in relation to workplace and education locations. There is, therefore, a strong argument for a reform of the whole current taxation system so that it is complementary to sustainable development objectives and so ensures a smooth integration of LETs systems onto that platform.

Perhaps understandably, of the three categories, it is the 'Polluter pays' type of mechanism that the EC is particularly keen to see introduced – as stated in the report, *Towards Fair and Efficient Pricing in Transport* (CEC, 1995). This is because the theory behind it is that polluters – be they car drivers, industrial companies or whatever – must pay a tax for the privilege of polluting. The message is clear – pollute and pay, or change your behaviour to pollute less and pay correspondingly less.

Spreading the burden (equity, income distribution and social inclusion)

For the remainder of LETs, notably the majority of those used across the United States, the major principle behind adopting particular revenue sources has been to raise as much money in as low profile and uncontroversial a way as possible. In general the rationale for public transport subsidized by these LETs is that of social inclusion. In consequence one might expect the sources of finance to be progressive, but in reality, the sources of income appear to be more pragmatic than adhering to a social basis of design. Although the expenditure is justified on equity grounds, the issue of whether the taxes and charges to pay for it are themselves equitable does not seem to feature in any significant way. For example, the general sales tax, the most widely used earmarked charge in the United States for funding public transport, is regressive. It falls disproportionately on the poorer in society and does not discourage social harm. However, given the legendary antipathy of US citizens towards paying tax, especially as local referendums are often required to introduce them, such apparently contradictory pragmatism can perhaps be understood.

The three systems

The evolution of local earmarked taxes and charges means that they bear a variable relationship to principles of public finance and the growing issue that the taxation system should be used not only to raise revenue, but steer the economy in the direction of sustainability and ecological efficiency. There are, not surprisingly, tensions between the rationales behind the three categories of LETs. The beneficiary pays and polluter pays LETs can appear to produce diametrically opposing signals. For example, the *Versement Transport* employer tax in France, where benefiting companies pay taxes towards new public transport infrastructure, may have the effect of persuading companies to relocate to areas less well served by public transport where the tax burden is reduced. Obviously as a result of this, one could expect an increase in car use among such employees, and a decline in public transport use – a very negative signal if the overall objective is to encourage people to use public transport.

By contrast, with the polluter pays mechanism, the signal operates the other way. For example, a parking charge hypothecated to public transport improvements both discourages car use and incentivizes public transport use – a double dividend. However, were such a polluter pays LET applied in a particular city, it could also lead to counter-productive relocational rebound effects.

This raises serious questions about the role of LETs, both in terms of them being effective policy mechanisms and also their relationship to principles of public finance. Indeed, can local earmarked taxes and charges even fulfil the basic requirement to be a reliable source of income for public transport? The following three Chapters explore these questions by drawing upon the findings of a CEC study entitled *Fair and Efficient Pricing in Transport – The Role of Charges and Taxes* (Van den Branden *et al.*, 2000). The wider policy and theoretical implications are then revisited in Chapters 6 and 7 of this book.

Van den Branden *et al.* (2000) developed the threefold 'polluter pays', 'beneficiary pays', and 'spreading the burden' into a total of categories and used this to analyse practical experience of a wide variety of implemented and proposed schemes from all over the world. These expanded categories were:

1   Employer/employee taxes
2   Property-related taxes
3   Development levies
4   Parking charges and fines
5   Charges for the use of road space
6   Local motor taxes
7   Consumption taxes
8   Cross-utility financing
9   Other miscellaneous LETs

The first three are based on the beneficiary pays principle – employment taxes, property-based taxes, and developer levies. These are examined in more detail in Chapter 3. Further details of the three polluter pays type mechanisms – parking levies, vehicle-related charges, and road user charges are provided in Chapter 4. The final three mechanisms – consumption taxes, cross-utility financing, and other miscellaneous LETs – the 'spread the burden' examples – are reported in Chapter 5. Finally, the above categorizations return in Chapter 6, which draws on the key examples detailed in Chapters 3–5 as well as others from the CEC project reported in Van den Branden *et al.* (2000).

Notes

1   Usually called *Transit* in the USA.
2   This argument ignores the so-called 'Mohring effect', which is the reverse of congestion, and reflects the positive externality that public transport users create for each other through the increased frequency that is (in the long run) associated with increased usage (Mohring, 1972).
3   Based on The Commission on Taxation and Citizenship (2000).

Chapter 3

Beneficiary pays: Economically-justified LETs

The beneficiary pays principle

As noted in Chapter 2, one of the oldest rationales for taxation derives from the concept of public goods that cannot be provided on a market exchange basis. A very large number of LETs have been developed in many nations to charge people and organizations for the collective benefits of public transport. This could involve a local charge to the area where public transport investment takes place or to employers located in that area. Whelan (2003) in his assessment of possible funding mechanisms for London identifies a large number of beneficiary pays LETs. Although this study is based on the beneficiary pays principle it does include a wide range of LETs that in this book are classified otherwise (e.g. road charging, sales and gambling taxes).

This Chapter contains a number of examples of LETs that have developed from the beneficiary pays approach. These include LETs on employment, on property, on land values, and on developers. Of these, the last is most common, with payments from developers being required to compensate for the transport impacts generated.

Since taxation to provide public goods is a well established rationale, LETs based on this principle are widely understood and generally accepted, even if the specific application of that tax is not. For example, while many an argument has centred on how high the rates of income tax should be, far less attention has been directed at how the system might be replaced. Consequently, Beneficiary Pays tax mechanisms are attractive to policy-makers as the public acceptance and legal problems are reduced.

However, as mentioned in Chapter 2, there can be problems in that a beneficiary tax to pay for public transport actually penalizes employers, property owners, and developers locating in an area with good public transport. This means that there is a risk that firms, residents, and developers may be encouraged to relocate in a site with no public transport but with no extra taxes either – clearly not a desirable outcome. Despite this, the rationale underpinning Beneficiary Pays LETs is fundamentally an economic one, with the argument being that improved public transport increases the level of economic activity in an area.

This Chapter reports on a number of examples where locally-applied

Beneficiary Pays taxes and charges have been dedicated to pay for local public transport facilities. In all, three variations of public transport-funding Beneficiary Pays tax were identified. These are levied on employment, property and on property developers.

Employment taxes

Despite the almost universal use of employment tax (i.e. income tax) to collect general revenues at a national level (and in many countries at a regional or state level as well), instances where money from an employment tax is earmarked to fund public transport systems are comparatively rare. Where they occur such LETs do not usually take the form of an hypothecated local income tax. The LET is upon employers rather than employees. This appears to reflect the benefit employers derive from improved public transport services and LETs are commonly based on the total company payroll.

Most local income taxes have a flat rate and they are also horizontally equitable as individuals of comparable incomes tend to pay comparable taxes. But, where the tax is not uniformly levied across a region, inequalities arise. Where a local income tax is higher in a city centre this could encourage the better off to live in the suburbs. Such problems can be circumvented if payroll taxes are adopted i.e. where the tax is based on the total salary at a place of employment rather than the place of residence. However, employment taxes can be controversial because commuters who live outside the taxation district in which they work have no say in opposing the tax. Employment taxes also encourage companies to relocate in the suburbs that may not be served by public transport, thus possibly increasing car use among its employees and visitors.

Employment taxes are not as stable as consumption taxes, such as LETs on sales or motor fuels, as they are more susceptible to economic conditions. Receipts are reduced in a recession, which is typically when tax revenues are most needed.

This section examines how employment taxes have been used in Austria, France and the United States. It also briefly reviews experience of student fees being hypothecated to fund improved public transport access to college campuses in the US, which although not strictly 'employment' taxes are reasonably closely related.

Dienstgeberabgabe employer tax Vienna, Austria[1]

While perhaps the best known example of an employer tax is the so-called *Versement Transport* in France (see later), this was actually preceded by the *Dienstgeberabgabe* in Vienna. This was first introduced on 1 January 1970,

Figure 3.1 A new section of the Vienna Metro. The cost of the Vienna Metro was partially met by the Dienstgeberabgabe Employer Tax

and was designed to pay for the construction of the city's underground railway.

Employers in the city with ten or more people must pay €1.1[2] per week per employee by the fifteenth of the following month, and must declare by 31 March the amount they have had to pay for the previous year. Employers must transfer money into the account of the city council, and where an employee is employed for only a part-week (say at the beginning or end of his/her contract), the full week fee must be paid. Exempted employees include: public sector workers (in government, electricity, public transport, telephone, post office etc.), employees aged over 55, apprentices, carers for the disabled, National Service conscripts, charities, staff on certified unpaid leave (e.g. people on maternity leave), workers employed for less than ten hours a week, and caretakers.

Where there is a high turnover of staff, employers can negotiate with the city to simplify administrative procedures. If the law is contravened, there is a penalty of €440, while if the city is defrauded the penalty is a €22,000 fine or six week prison sentence. Overall, the tax raises about €21.5m a year, which covers 10% of the overall construction budget of €219m a year. No major problems have been reported in implementing the levy, but despite this such a charge has not been levied in other Austrian cities.

Le Versement Transport employer tax, France

One of the most established and widely used LETs is *Le Versement Transport* (VT) employer tax in France. First levied in Paris in 1971, the VT is now collected in all urban areas with more than 100,000 population, and in 80% of cities with populations between 20,000 and 100,000 (Farrell, 1999*a*). The tax is levied on employers with ten or more employees, whose place of work is situated within a specified 'urban transport radius'. The only employers whose contribution is reimbursed are those that can demonstrate that they provide housing for their workforce at the workplace or provide transport by 'collective' means for all or some of their employees. The transport of this staff must be free (laws No. 73-640 of 11 July 1973 and No. 8-52 of 2 January 1985). Also included are employers who employ staff within the conurbation of the new towns or in certain industrial and commercial zones (law No. 75-580 of 5 July 1975, article 2).

The VT is paid as a percentage of the employer's total payroll costs, and

Figure 3.2 The Carte Orange public transport travel card in Paris is subsidized through employer contributions

paid to the limit of the ceiling imposed for social security to the Union pour le Recouvrement de la Sécurité Sociale et des Allocations Familiales (URSSAF, or the body responsible for social security repayments). For every payroll cost there is a maximum levy rate, which is decided by the local authorities on the basis of state dictated limits (see below). The local transport authorities alone decide how to spend VT monies in their areas, bearing in mind that VT must be spent on some form of public transport (Copenhagen Transport *et al.*, 1995; Hass-Klau and Crampton, 1999; Farrell, 1999*a*). In addition, in the Paris region, employers have been required to refund half the cost of the *Carte Oranges* season ticket since 1983 (Transport for London, 2000).

Three-quarters of the eligible authorities set VT at its maximum allowable rate. These rates are (as a percentage of the wages bill): Ile-de-France: central areas 2.2%, inner ring 1.6%, outer ring 1.3%; provincial cities >100,000 population: with fixed track system 1.75%, other 1.0%. Provincial cities <100,000 population 0.55%. VT is used to fund capital investment in public transport, extend or improve services, or subsidize fares (Farrell, 1999*a*).

The VT was introduced because of a growing realization in the 1960s amongst local French politicians that traffic congestion was becoming a serious problem, and that the financial needs of public transport had been neglected for too long (Meyer, 1996). As a result, Finance Law No. 71-559 of 12 July 1971 was enacted to establish a new fund to address this, by raising money from additional revenues from police fines (a polluter, or rather offender, pays LET considered in Chapter 4), and from employers. Thus the VT contribution was introduced in Paris in 1971, and in other cities with populations of more than 300,000 in 1973, 100,000 in 1974, 30,000 in 1982, and more than 20,000 in 1992. The original purpose of the VT was to encourage commuters to travel by public transport by discounting fares. The subsequent loss in revenue was then recovered through a tax on employers. But, following a law of 4 August 1982, relating to the participation of employers in public transport finance, the levy may also be used for general fare subsidies (Copenhagen Transport *et al.*, 1995).

The VT has provided significant sums for public transport investment. It has, for example, enabled the modernization of the Paris metro and the construction of metros in Lille, Lyon and Marseille, along with several new tram systems. In Paris in 1998, VT raised €710m,[3] or 24% of the cash in-flows in Paris. Of this, one-third is allocated to support depreciation costs while the remainder subsidizes fares. The rest of the system's operational cost is covered by fares (€1.4bn, including the employer fare subsidy for the *Carte Oranges*) direct public subsidy (€580m), commercial activity (€240m), and concessionary fare support (€90m) (Transport for London, 2000). But there have been problems. From the mid-1980s the VT started to fall behind as a funding source due to the methods of calculating VT, the growth in unemployment, and because some firms relocated due to the differences in the rates of taxation between Paris and the outer ring (Ridley and Fawkner, 1987).

Thus, while the VT stimulated local authority investment in modernization and expansion, it is possible that this development was carried out in a state of euphoria created by the availability of 'easy' finance, without proper evaluation of its medium term consequences. In addition, the charge has been affected by the increase in unemployment, which has slowed down the rate of growth in its revenues (Coindet, 1994). The progressive reduction in the proportion of expenditure covered by income is disturbing, and there is a risk that relaxation of the rules will divert all tax revenue to cover operating costs, with no margin for investment. It now appears that VT has played out its role as the driver of development, and that it will not finance further major construction work unless ways are found to increase its revenue yield. Significantly, transport practitioners in the Ile-de-France are now looking at adopting other mechanisms (such as a form of workplace parking levy – see Chapter 4) to supplement the VT.

Employer taxes in Portland and Eugene, Oregon, United States[4]

An example of a direct payroll taxes LET is provided in Portland in a system that also predates France's *Le Versement*. Payroll taxes to support local public transport have been imposed for over 30 years by the Tri-County Metropolitan Transportation Authority (Tri-Met), and in Eugene by the Lane County Mass Transit District. The State legislature permits the district to adjust the tax rate providing it does not exceed the statutory ceiling of 0.6%, and both Portland and Eugene levy this maximum rate. The tax was introduced on employers in Portland from 1970, while the self-employed have been required to pay the tax since 1982. Taxes are paid quarterly, by employers (including self-employed) within the transit districts. By law, government employees are exempt from paying the tax, while the State of Oregon government pays an amount in lieu of the tax on the payroll of its employees working in the district. The State Department of Revenues collects and administers the tax. All revenues, after handling costs incurred by the State are deducted, are forwarded to the transit district.

In the 2001 financial year, the payroll tax generated a net of $US151.6m (€170m), or 56% of Tri-Met's operating budget.[5] Passenger revenues contributed $US51.7m (€57.9m), other sources $US59m (€66m), interest $US8.4m (€9.4m), and the State cigarette tax $US1.5m (€1.7m) (Tri-Met, 2001).

Employment tax experience elsewhere

Elsewhere in the United States, dedicated employment taxes can be found in the States of Kentucky, Ohio, Indiana and Washington.

In Louisville and Jefferson County, Kentucky a 0.2% 'occupational tax' was approved by voters in November 1974 and introduced on 1 January the following year to provide capital and operational matching funds for the Transit Authority of River City (TARC). Revenue from the levy is paid into the Mass Transit Trust Fund (MTTF). The tax levy covers 73% of operating costs of the transit system, and generated around $US32m (€36m) for the MTTF in 2000. One consequence is that the TARC base fare of $US0.75 (€0.84) is one of the lowest in the US (City of Louisville, 2000; Pattison, 1999). The city is currently bidding for Federal money to build a light rail line. It is proposed that the 10% local share of the money required would be raised through a 0.05% increase in the occupational tax levy. Also in Kentucky, the districts of Boone, Kenton and Campbell use payroll taxes to fund Transit Authority of Northern Kentucky (southern Cincinnati) (Goldman *et al.*, 2001).

In Cincinnati, Ohio, a 0.3% income tax on all working in the city provides funds for the Southwest Ohio Regional Transit Authority (SORTA). The tax was established in 1973, when voters approved a tax increase to buy the Cincinnati Transit bus system. This led to the establishment of Metro, the bus operating division of SORTA, which serves an area of 867,000 people. In 1997, the levy funded 54% of the system's operating costs i.e. $US30.3m (€34m) (SORTA, undated; Goldman *et al.*, 2001; Pattison, 1999).

Lafayette and South Bend transit corporations in Indiana are part funded by dedicated local income tax revenues. Lake County is also authorized to adopt employment tax of up to $US1 (€1.1) per employee per month, with half paid by employers, and half by employees (Goldman *et al.*, 2001). This is a relatively small amount, and could be viewed more as a 'Spreading the Burden' LET than an example of the Beneficiary Pays principle.

The State of Washington imposes a number of business and occupation taxes/business licence fees, which are based on gross proceeds, business type, staff numbers, floor area etc. While these are often general taxes, in thirty cities the funds raised are dedicated to fund measures to help cut peak-hour congestion. For example, the employer tax charges $US2 (€2.2) per employee per month for 'high capacity transportation' projects or HOV lanes, commuter rail, or vanpools. Exemptions from the tax are offered to employers who participate in Traffic Demand Management (Mobility Management) schemes (Goldman *et al.*, 2001).

Student surcharge, University of California, Berkeley, California[6]

At the University of California, Berkeley, local transit operator AC Transit sought to raise more revenue and introduce a cheaper and improved universal pass programme for students. This was achieved in collaboration with the University, through a subsidy for the class pass based on a surcharge on

Figure 3.3 A surcharge
on student fees helps pay
for public transport at the
University of California's
Berkeley Campus

student registration/tuition fees. The surcharge costs students less than a monthly adult pass ($US45 (€50)/month). Instead, all Berkeley students were subject to a surcharge on student registration fees of $US18 (€20) per semester, from which AC Transit obtains $US10 (€11), University of California, Berkeley's Parking and Transportation Operation $US2 (€2.2), and University of California financial aid programme $US6 (€6.7).

The measure was voted on in a special student referendum, and this resulted in 88% voting in favour. As a result, the fee was introduced early in 1999 (AC Transit, 1999). Altogether, AC Transit receives roughly $US6m (€6.7m) per year for operating dollars (based on 30,000 students). In addition, AC Transit receives a payment (to be determined) from University of California, Berkeley for each class pass holding student using the TransBay service.

Student surcharge, University of Wisconsin-Milwaukee and Marquette University, Milwaukee, Wisconsin

In an arrangement similar to that in Berkeley, Milwaukee County Transit

(MCTS) developed partnerships with the University of Wisconsin-Milwaukee and Marquette University whereby the universities pay $US29 (€32) per student per semester to MCTS and all students who wish to use the services ride free. The programme is called UPASS. The revenue impact is minimal, though receiving the revenue up-front from the universities is helpful, and use of the system by more students is expected to generate additional regular fare trips. In essence, MCTS found a way to increase ridership, ease parking and congestion in and around the universities, increase mobility for students, and develop stronger community relations without additional public funding (CUTR, 1996).

Employment taxes: conclusions

Overall, there are a number of ways to provide earmarked employment taxes to fund public transport. They provide significant revenues at a very low percentage charge rate, with a highly stable tax base and revenue stream. However, revenues from the tax tend to drop significantly during economic slow downs – which is usually when public transport subsidies are most needed.

Employment taxes are also progressive and thus relatively equitable, and there is a direct cost/benefit relationship when using the revenue to improve public transport, as employers gain from increased transport efficiency and a wider labour pool. However, there is a danger that employment taxes could push firms to relocate to less public transport accessible areas to escape the tax. If the revenues are efficiently spent this should not be a problem, as the value to employers will outweigh the costs. Relocation is most likely if funds are used inefficiently and do not produce sufficient benefits to employers and their staff. This may happen, for example, if LETs revenues are used to fund expensive and poorly patronized prestige projects or to cut fares for non-employees.

Property taxes

As with the employer tax, part of the logic behind a property tax is the concept that public transport is a 'public good'. By providing a public transport service, the inhabitants of the properties served benefit, in this case by an increase in the value of the property and land on which it is built. Thus, the tax is a means of recapturing part of the value enhancement that public transport provides. Value capture is a mechanism 'by which the agency responsible for the development of the urban transport infrastructure captures part of the financial benefit gained by land developers or the community at large' (Tsukada and Kuranami, 1994). This benefit is reflected in an increase in the real property values, '. . . reflecting . . . improved accessibility and an increase in business opportunities'. This process of 'value recapturing' or 'realizing betterment', can be divided into taxes paid regularly to the local or regional government, which then earmarks a specified amount to subsidize public transport, and the usually 'one-off' or irregular developer levies. Developer levies will be looked at later in the chapter.

Paying for the provision of public services through the collection of local property (or land) taxes, is a fairly common method world-wide, being evident throughout Europe, Australasia and North America. The reason property taxes are thought appropriate as a source of revenue for local governments is the connection between the types of services funded at the local level and the benefit to property values (Slack, 2001). For the most part such monies are collected by local authorities and allocated to each sector according to the prevailing political objectives, but in the United States, there are examples of directly earmarked general property taxes. Indeed, until the recent increase in the use of sales taxes, earmarked property taxes were the most common method for paying for public transport systems. Examples have been found in several cities including Anchorage, Minneapolis/St Paul, New York, Denver, Detroit, Milwaukee and Miami (Simpson, 1994). In addition a mortgage tax, effectively a form of property tax, is used to fund public transport in several parts of New York State, including Albany and Buffalo (Bushell, 1994).

Property taxes became popular among local government officials because they are based on immobile objects and so are difficult to evade. In addition, they are easy to administer because local authorities administer land ownership records too.

Property taxes are charged to property owners as a percentage of the current assessed value of property. There are two main ways localities use property taxes to fund public transport projects. The first is to earmark a specific portion of annual revenues, which is rare. The second is to direct a property tax increase or surcharge, temporary or permanent, to a specific purpose. Typically, property taxes are implemented using a three-stage process. First, a tax assessor estimates value of land and buildings in each parcel. There may be periodic calculation or the value may be frozen at a point in time. Then the assessed value is assigned to the property, depending on use of land. Finally, the taxation office (separate from assessment office) sets a tax or 'millage rate' by dividing the local government's total budget for the upcoming year by the total assessed valuation for the area. An individual parcel of land may face different millage rates for each governmental entity serving it, including city and county governments fire districts etc. A parcel's property tax is the product of its assessed value and the sum of all applicable millage rates (Goldman *et al.*, 2001).

General property tax, Minneapolis, Minnesota

One fairly typical example of how the property tax works in the USA is Minneapolis, Minnesota (Gibbons, 2000; Metropolitan Council, 1999). In

1999, the dedicated property tax there raised $US62.5m (€68.7m), covering just over 40% of Metro-Transit's $US156.2m (€175m) operating budget, while fares contributed 33%, the State of Minnesota 19%, and Federal funds 3%, with contract revenue and other sources providing the balance. The local property tax was introduced in 1971, when the Metro Transit Commission was formed to operate public transport, and currently applies to the 970,000 or so residential properties in the Metropolitan Council area. The tax is a flat rate in each county, but is 'feathered' so that residents of the counties better served by transit – such as the downtown areas – pay more than suburban householders. The transit portion of the tax is collected annually by each county, and then forwarded to the Transit operating division. At the present time there are significant discussions to what the future funding requirements are, and how these can best be served.

There are several problems with the property tax as it stands. Firstly, the property tax is capped so that revenue can only grow when property values rise. This means that for several years not enough money has been raised, requiring the Commission to request additional funding from the State. Secondly, and related to this, property prices, and thus the levels of service have risen in the 'cash rich' suburban counties, but not in the city area – where 90% of services are provided. Primarily as a result of these factors, it may well be that in the future a dedicated local sales tax is introduced to replace the transit element of the property tax.

More broadly, property taxes are unpopular with tax payers because they are paid in lump sums rather than incrementally. They are also used for services such as schools that are used by a limited segment of the population (as is public transport). Often, their administration appears arbitrary, and the ultimate tax bills appear to bear no relation to the household's income or ability to pay (Goldman et al., 2001).

Dedicated regional property taxes, Vancouver, British Columbia, Canada

A non-residential property tax in the Greater Vancouver Transportation Authority Area was estimated to have raised $C35.5m (€25m) for transit operations in 1999/2000.[7] The rate of tax varies according to the class of utility and industry. Based on the 1998 rates approved by the Vancouver Regional Transit Commission, Class 2 utilities pay 0.1376% of the assessed value of the company; Class 4 major industry and Class 5 light industry pay 0.13381%; and Class 6 business and other pay 0.09627% (GVTA, 1999).

The Greater Vancouver Transportation Authority or 'TransLink' (which has responsibility for roads as well as transit) began operating on 1 April 1999, after being established by the GVTA Act. Altogether, 63% of the $C633.1m (€45m) 2002 annual budget was derived from dedicated local or regional taxes

(TransLink, 2002*a*). Also, in Vancouver, a hospital/transport levy on residential properties of \$C52.6m (€37.3m) – equal to the 1998 requisition for the Greater Vancouver Regional Hospital District (GVRHD) – has been transferred to the Greater Vancouver Transportation Authority under the GVTA Act. In addition, the levy includes \$C1.9m (€1.3m) budgeted expenditures for the Transportation Planning and Hospital Planning Departments of the GVRHD which has since been dissolved (GVTA, 1999). Altogether, the 2001 Budget revealed that dedicated property taxes raised \$C93m (€66m) for the GVTA, which averaged out at \$C59 (€42) per household a year. This amounted to 33% of the GVTA's income (TransLink, 2002*b*).

Benefit assessment districts

To overcome some of the problems associated with the general property tax the so-called '*Benefit Assessment District*' has been used in the United States. *Benefit Assessment* (also known as a *Special Assessment* or a *Local Improvement Charge*) is a charge on property used to pay part or all of the cost of capital improvements that enhance the value of the property. These capital improvements can include public transport.

A problem with any general property tax is that it can be levied without respect to the benefit to the land taxed. But Benefit Assessments must be proportional to the benefit conferred upon the land as a result of the improvements. In fact, Benefit Assessment Districts cannot be established unless an engineering report can identify, and provide a method to calculate, the special benefits. A general property tax need not satisfy this requirement (Knox, 1996). In addition, assessment law requires that property owners be given notice and provided with a public hearing before an Assessment District can be formed. Finally, the protest procedure common to assessments allows property owners the opportunity to block imposition of the assessment.

In theory, assessments should apportion the project capital costs to benefiting property owners based on the value of the additional benefits received by each property. In reality, though, it is difficult to isolate the impact of one capital expenditure from other influences on property values. Overall, assessments are not as efficient as user fees because the charge is not directly related to the use of the service, although they do approximate benefit taxes more closely than the property tax does (Slack, 2001).

One of the few longer running examples of a Benefit Assessment District is in San Francisco, where the tax system has been used to raise money to build and operate the Bay Area Rapid Transit (BART) system.

General obligation bonds to fund BART, San Francisco, California[8]

On 6 November 1962, the citizens of the San Francisco Bay Area Rapid

Transit Region (i.e. the City of San Francisco, and Alameda and Contra Costa Counties) passed a local resolution to allow money to be raised to pay for the 114 km BART rapid transit system. To do this, BART issued General Obligation bonds (incurring debt), that were to be repaid using money hypothecated from property assessments throughout the three counties.

The original construction of BART, which began operating in 1972, was financed by the sale of $US792m (€870m) in General Obligation bonds while the cost of the annual debt service was $US45m (€49m). Other LET sources were added to fund BART. The Bonds were supplemented by San Francisco-Oakland Bay Bridge toll revenues, and in 1969 voters approved a 0.5% permanent sales tax to finance construction bonds. There was also a final LET in 1999, when property owners paid 16.7 cents per $1000 (16.7 Euro cents per €1000) of assessed value to pay off the final instalment of the initial debt incurred to build BART. This tax is now gone. Using these sources, in 1999, BART repaid the principal and interest of its construction debt.

The main advantages of the use of Benefit Assessments is that they provide a stable revenue stream that yields a high bond rating and hence low interest costs. It is also somewhat progressive, and is simple to operate.

Although the initial General Obligation Bonds for BART have been paid off, they are set to return. A vote in November 2002 sought to authorize the sale of bonds to the value of $US1.05bn (€1.16bn) specifically to strengthen, seismically retrofit, improve, and replace BART facilities. If the bonds are approved, the District expects to sell them in four series over time. Principal and interest on the bonds would be payable from the proceeds of tax levies made upon the taxable property in the District. This is expected to be levied at the rate of between 3.3 cents and 14.2 cents per $US1000 (3.3–14.2 Euro cents per €1000) of assessed valuation over a 40-year period. In practice, the District estimates that the highest estimated annual tax for these bonds for the owner of a home with a net assessed value of $US300,000 (€330,000) would amount to approximately $US42.50 (€46.75).

Benefit assessment districts programme, Los Angeles, California[9]

A similar benefit assessment scheme to that which supports BART has been operating to help fund part of the Los Angeles Red Line since July 1985 when the Southern California Rapid Transit District (RTD), one of the predecessor agencies for the Los Angeles County Metropolitan Transportation Authority (MTA), formed two Benefit Assessment Districts.

Assessments received from these districts are used to pay off bonds issued to pay a portion of the station construction costs of the first segment of the Metro Red Line. Assessment payments will terminate in 2008–2009. Overall, 9% ($US130m) (€146m) of the $US1.418bn (€1.588bn) cost of the Metro Red

Figure 3.4 Special assessments helped build BART . . .

Line between Union Station–Westlake/MacArthur Park was raised through assessments.

In Los Angeles, the annual assessment rate is determined by dividing the annual bond repayment by the assessable square footage and factoring in the last three years' delinquency rates. The assessment rate is calculated on an annual basis. It is levied either on the gross area of the assessable improvement or the parcel area, whichever is greater. Properties that are subject to the assessment include assessable improvements and assessable parcels.

Assessable improvements include offices, retail stores, hotels/motels, and other commercial properties. Assessable parcels with non-assessable improvements include wholesale, manufacturing, industrial, improvements vacant due to regulatory code, parking, and vacant land. Finally, exempt properties include residential, non-profit owned and used, and publicly-owned and used land.

The 2000–2001 rate for District A1 is $US0.218 per assessable square foot (€2.35 per square metre). District A1 covers 1,205 acres (4.88 square km) and encompasses the downtown area of the City of Los Angeles. This area includes Bunker Hill, the Civic Center portions of Chinatown, Little Tokyo and the Financial District, and includes four Metro Red Line stations, which opened in 1993. District A1 contains 2,676 properties, of which 1,254 are assessable, and in 2000, contained 63,238,725 assessable square feet (5.88 square km). Boundaries are set at a one-half mile distance (800 m) from the stations, and $US123.7m (€139m) was generated through bond sales which was applied toward station construction costs.

The 2000–2001 rate for the second Benefit Assessment District – District A2 – is $US0.273 per assessable square foot (€3.29 per square metre). District A2 covers approximately 207 acres (0.83 square km) and includes Westlake/ MacArthur Park. It is located on Wilshire Boulevard, midway between Miracle Mile to the west and the Central Business District to the east in the City of Los Angeles, and its boundaries are set at a distance one-third mile (about 530 m) from the station. Around $US6.5m (€7.3m) was generated through bond sales which was applied toward building one station which opened in 1993. Approximately 456 properties are located within District A2, of which 233 are assessable. In 2000, District A2 contained 3,291,084 assessable square feet (0.31 square km).

Japanese experience of earmarked property taxes[10]

In Japan, the legal system is used to empower government agencies to collect taxes from land developers, residents and businesses directly or indirectly benefiting from a railway project. These tax revenues are then used to establish a special Railway Development Fund. This approach has been successfully applied by local governments in Japan to provide direct subsidies for railway construction or operating costs, or low interest loans to railway enterprises.

Central government has chosen to subsidize the construction of underground rail lines operated publicly or by public/private joint ventures. These are often in the form of a corporation whose stock is jointly owned by local government, industries involved in, or benefiting from, the project in question, and leading companies in the region such as banks and power companies. In 1990 the government subsidy covered 70% of eligible construction expenses, equivalent to 60% of total construction costs, payable in ten-year instalments beginning once the service has started. One condition is that central government will provide half of the subsidy if local government provides an equivalent amount.

To secure the resources for this shared burden, some local governments – such as Sapporo, Sendai, Fukoka and Kitakyushu – have introduced a Special Railway Fund, financed principally by earmarking the incremental revenue from increases in existing local taxes. This can be regarded as a form of value capture when the incremental revenues come from an increase in corporation tax, providing the business corporations are the direct beneficiaries of improvements in railway services. When the Sendai municipal government, for example, established the Sendai Municipal Rapid Mass Transit Construction Fund in 1980 this was funded by a 14.5% increase in local corporation and business establishment taxes. The revenues were then used to subsidize part of the Y240bn (€2.18bn) construction and interest expenses required for developing new railways lines in the city.[11]

In Osaka, the municipal government increased existing local property taxes, and earmarked the revenues to fund an urban rail system. All the landowners and leaseholders within a targeted area collectively contributed a part of the system construction cost, in a scheme resembling US Special Assessment Districts. The area benefiting from the system was defined on the basis of distance from stations, ranging from 360 m to 720 m. In addition, different tax multipliers were used for four categories of location, based on distance from the city centre. Taxpayers contribute according to the area of land they own, the multiplier, and a formula including distance to the nearest station.

When the Kobe municipal subway was extended to Suma new town, the local government set up administrative guidelines to obtain contributions from the new town developers for the railways. The new town developers were required to give land to the railway for free and pay for all the railway construction costs. The fees were charged according to Special Assessment Districts set up around the rail stations by the municipal government. The specific development charges were determined by a formula that mainly considered the distance of the development from the station. The developers were also expected to make large-scale developments in the areas around the stations. This system is unusual in that such a heavy burden was placed on developers (Bell, 1993).

Seventh rail corridor, Mumbai, India

A location benefit charge has also been proposed for part-funding the 'seventh rail corridor' in Mumbai (Bombay). Employers and residences in the vicinity of the new metro stations are to contribute, as they are expected to gain from the increase in land values. However, because the low rateable value base of property rates in Mumbai, this source is not likely to raise more than RS100m (€2.3m) (Dalvi and Pantakar, 1999).[12] Real estate development surpluses for the seventh rail corridor in Mumbai are expected to be in the order of RS63bn (€1.47bn).

Other examples

A system based on specific improvement assessments was used to part finance the construction of 35 km of a railway line in Milan, Italy. The tax based on the increases arising in values of developed land within 500 m of a station, raised €18.6m.[13] However, the mechanism has now been replaced by a general tax on transfers of property that is not earmarked to public transport. The Barcelona Entidad Metropolitana de Transporte in Spain raises over €12m a year from property taxes to help cover both operating deficits and investment costs (Farrell, 1999b).[14] In Lisbon, Portugal, a special tax on residents benefiting from the construction of a new rail extension to Benfica was considered, but in the end the line was funded through a government grant (Ridley and Fawkner, 1987).

Land value increment taxes

While not strictly a property tax, land value increment taxes (land value capture taxes, betterment taxes or valorization taxes) are levied to capture the increase in land value generated by public investment rather than the actions of the landowner. The unearned increments can be captured indirectly through conversion into taxes or fees, or directly through on-site improvements that benefit the community at large.

In order to be successful, there are a number of elements that must be taken into account. First, the tax must be linked directly to the benefit (Slack, 2001). Second, the timing of the project is of the utmost importance. Specifically, problems can be encountered if development slows due to an economic recession. And, if a new metro system is being built it must be ready when the new buildings open for business, but should not be too long before that happens (Jensen, 2002).

In general, land value taxes are collected in large amounts from a small number of taxpayers and so are easier to administer than property taxes.

Capturing betterment to fund the Ørestadsbanen, Copenhagen, Denmark[15]

One example of using the land value tax to capture benefits arising from improving infrastructure is the Ørestadsbanen automated light rail system project in Denmark. Here, various plots of land situated in the Ørestad area, a new town near central Copenhagen, are being developed and provided with a light rail system. This will be financed by realizing the actual increase in the value of land that the light rail system will generate, with the Danish state and the City of Copenhagen providing a guarantee until the money can be realized (Copenhagen Transport et al., 1995; Ahm, 1999).

The three-line mini-metro is being funded by the sale of a vacant long, thin 320-hectare site in the Ørestad area close to the city centre. The ownership of the Ørestad area was transferred in 1993 from the joint ownership of the City of Copenhagen and Danish Government to a new development agency called Ørestadsselskabet (OS). Initially the metro will be funded through loans, which will be redeemed with the income from land sales (around two-

Figure 3.5 The Ørestadsbanen automated light rail system

thirds) and the proceeds from running the metro (approximately one-third). Forty five per cent of the company's funding requirement will be covered by a loan from the Ministry of Finance, while the remainder will be covered in the domestic or financial markets.

Property taxes: conclusions

The main advantages of earmarking property taxes are that most local governments have administrative systems in place for assessing real estate values and collecting taxes on this, which reduces administrative costs. Property taxes also provide a relatively large and stable revenue base. The limitations are that the tax is very visible in that it is paid directly in periodic lump sums. This means taxpayers are far more aware of how much tax they are paying. While this increases public accountability, it also raises the visibility of the tax and thus increases taxpayer resistance (Slack, 2001). A further problem is that the base of the property tax may not increase because tax authorities rarely update property valuations on an annual basis. Therefore to maintain property tax revenues in real terms it is necessary to increase the rate of the tax. Finally, property taxes can be unpopular because they are used for services that are used by a limited segment of the population, and are somewhat regressive as they bear no relation to the household's income or ability to pay.

Finally, a review of cases where property taxes had been used to finance transport projects (Farrell, 1999b), found that few projects produced good results. The lack of good results has more to do with the spending of the money than the means by which it was generated. However, if the revenue is not spent with efficiency this undermines acceptability of the

continued on page 72

continued from page 71

funding mechanism. It is telling that this was also true for employee taxes, and could be seen as a general feature of LETs.

  More specifically, Farrell also notes that crises in the property market reduced enthusiasm for this type of funding. She also identified that property development was been seen by banks as too risky to be used as security for infrastructure loans. The latter is important, and suggests that this form of LET may be limited to situations where exceptional rises in land values occur, possibly where land is already in state ownership or has very low acquisition costs.

Developer levies and charges

As part of providing permission for a development, it is common in most developed countries for the planning authority to require contributions to compensate the community for the extra costs of public facilities that are produced or required by that development. Paid at the time of planning permission or a building permit being granted, fees are placed in a fund designated for construction of certain types of facilities that can be linked to solving the problems that the development causes. This often includes paying for the provision of transport infrastructure, like a new road junction to serve the development site. A variation is developer financing, whereby the property developer finances the construction or expansion of a public transport facility in exchange for the right to build residential housing, commercial stores, and/or industrial facilities. For example they may provide bus stops or, for a large development, a rail station. The private developer may even operate the facility under the oversight of the local government. Developer financing arrangements are called a lot of things, including capacity credits, impact fees, exactions, or development gain. However, developer financing is almost always limited to certain locations, particularly areas of rapid growth. The developers may not like to pay or manage the facilities required and often resist, even to the point of engaging in litigation with local authorities (USEPA, 1999).

  In many respects, developer levies are attractive to local government. They appear to pass part of the costs of infrastructure and other service needs that are generated by new development directly onto those who profit from them. But high development fees and levies may push developers to move to places where there are lower fees or where they can influence the political process more easily. Such areas are likely to be on the metropolitan periphery, contributing to suburban sprawl. In fact, local residents opposed to growth may use this mechanism to prevent development (Teitz, 1999).

  Development levies can take a number of forms. These include:

◆   impact fees, whereby part of the cost of transport would be recovered by special charges on different land uses, usually levied at the time of new development of properties in the benefiting areas according to some type of 'impact formula';

- development charges, which are broadly the same as impact fees but which are negotiated on a case by case basis between the planning authority and the developer without recourse to a formula.

Other, non-LET methods that are not further described here include:

- benefit sharing, which is similar but which is tied specially to the increase in property values resulting from public investment;
- payment by the property owner for all or part of a line extension or station that is integrated into the development;
- sale of surplus land or air rights by the transit authority to developers to recover some of the authority's costs. In fact, in some cases the authority has purchased extra land for subsequent sale or development;
- connection fees, whereby a property owner pays a specific fee to be connected directly to the transit system.

Given that some sort of developer fee linked to the right to build on a site is widespread practice, the following examples concentrate on some more innovative uses of this mechanism.

Impact based developer fees in Cambridge, UK[16]

Unlike most local authorities in the UK that more typically rely on negotiated payments by the developer, Cambridge City Council has developed a transparent mechanism for charging developers according to the number of trips their site will generate. To do this, the council drew up two Area Transport Plans (ATP) – one for the Eastern Corridor (ECATP) and one for the Southern Corridor (SCATP) which were adopted by the Supplementary Planning Guidance to the Cambridge Local Plan in 2000.

Broadly, the purpose of the ATPs is, firstly, to identify what new transport infrastructure and service provision is needed to facilitate large-scale development in Cambridge. And, secondly, to implement a fair and robust means of calculating how individual development sites in the area should contribute towards the fulfilment of that transport infrastructure. The ATPs do this by designing and costing the transport system necessary to support the planned development. A fee per trip per day is then derived by dividing the cost of the transport upgrade by the number of newly developed trips in the ATP zone.

At present, any development within the relevant ATP area that generates more than 100 additional person trips (all modes) per day is liable for payments, which are charged on a per-trip basis. The ECATP review is proposing a contribution of £229 (€327) per additional generated trip[17] and may also result in a lower threshold of 50 person trips per day.

New town development charges in Japan

In Japan, the Ministry of Transport and the Ministry of Construction set up a system where connection charges and a site fee go towards funding rail construction. This is designed to improve private railways that connect the new town to the city centre commonly located 30 km–40 km away. The railway is built by a local public body and the Japan Public Railway Construction Corporation. The developers and landowners that own land where the private railway is going to operate must sell the portion of land that the railway needs at a reasonable price and pay for half of the railway construction costs associated with their previously owned property (Bell, 1993).

Public transport revenue subsidy, UK

As a rule, in the UK, developers provide a transport benefit as an agreed part of a project. This is easier to specify for infrastructure. For example, on the eastern edge of London, the Chafford Hundred railway station was built as part of the development of the large Chafford Hundred housing development. Another example was that the developer of Canary Wharf in the London Docklands, Olympia and York, contributed £400m (€570m) to the £2.76bn (€4.0bn) cost of extending the Jubilee Line from Green Park to Stratford in 1992.

Such planning consent levies are less easy to use to support revenue costs, such as subsidizing a bus service, than for spending on capital projects. Yet for new developments, revenue subsidies to public transport can be very important. One planning agreement that sought to provide this was one made in March 1997 (Moore, 2001) between Bracknell Forest Council and developer Helical Bar Developments (South East). Here, a contribution of £50,000 (€72,000) indexed from agreement to the date of payment was to be paid on occupation of the site by the developer. This was to cover the costs of providing public transport services whose routes include a link between the Western Industrial Area of Bracknell and the railway and bus stations in Bracknell town centre. The contribution had to be spent within five years of the contract date, otherwise it had to be repaid by the council. In this case, the developer paid £52,927 (€75,600) in December 1998, and the money was not spent as of summer 2001. More typically, contributions are earmarked to 'integrated transport' rather than public transport specifically.

Similarly, in the case of the Ocean Terminal by the Port of Edinburgh, the developer has made available 0.5m (€0.7m) to be paid to the City of Edinburgh Council for public transport improvements such as bus shelters and public transport information. It has also agreed to underwrite any shortfall of public transport operators not providing 5,000 seats per day (Mathie, 2001). This was agreed under a planning agreement (Section 75 of the Town and Country Planning Act in Scotland).

Voluntary exactions, Palace Quarter, Den Bosch, Netherlands

In Den Bosch in the Netherlands a public-private partnership is to develop industrial land to the west of the city's railway station. Once development starts, each developer (there are three each owning 25% of the land) has agreed to pay the City Council a fee for the land and the right to build on it. Of this, around 8% will be dedicated to a 'large works fund' that will pay for necessary off-site infrastructure including bus lanes (Rye, 2002).

Voluntary exactions, Nissei New Town, Japan

As their name suggests, voluntary exactions are different in that a pre-determined fee is not forced on the developers through laws or regulations. Developers and railway companies work together in making an official and biding financial arrangement that the developers should pay for a portion of the railway operating – not construction – costs. This type of exaction was negotiated between the Nose Railway Company and Nissei New Town, as well as between the Hokuso Development Railway Company and the Chiba Prefecture. In the case of Nose Railways, the developer also supplied half of the construction funds for an extension to Nissei new town and helped pay for an increase in the line's capacity (Bell, 1993).

Impact fees

Developer levies are a one-time charge. In general, impact fees differ from developer charges in that they are paid by a broader segment of the population. However, impact fees do not provide capital much in advance of development, unless impact 'rights' are sold up-front, and thus it may be hard for localities to ascertain capital needs and thus the size of fees. In the United States, impact fees are criticized for deterring development and increasing new housing costs, and resulting in competition between local authorities.

Relatively few cases of impact fees where money is dedicated to public transport were identified. This section looks at examples in San Francisco, Hamburg, and Toronto.

Transport Impact Development Fund, San Francisco, California, USA[18]

The Transport Impact Development Fund (TIDF) is a one-time fee designed, implemented and operated by the City and County of San Francisco to recover all incremental costs to the San Francisco Municipal Railway (Muni) from new office developments built in the TIDF assessment district. The fee is based on the floor space of the property ($US5 per square foot, or €60.3 per square metre) and assumes that the building will continue to impose costs on transit over an assumed 45-year lifetime. This is to cover the additional cost of providing transit services (additional rolling stock, services, personnel, fuel, electricity, facilities, and the maintenance, repair, replacement, and operation of the vehicles and facilities) caused by the new office development for that time. Payment (due on 50% occupancy of the net rental area or issuance of the first temporary permit or the final certificate of occupancy, whichever first) is from the developer to the city, which transfers it to Muni.

The fee came about because substantial downtown development in San Francisco in the late 1970s led to fears that the transit system would become overburdened, unless substantial investment was made in it. As there was insufficient general revenue for the required investment, and as there was fairly strong opposition among tax payers for other local taxes to fund transit, the San Francisco Public Utility Commission decided, in December 1978, to review legal aspects of a development impact fee. This resulted in the legislation to establish the Transit Impact Development Fee (TIDF) being enacted by the San Francisco Board of Supervisors in Chapter 38 of the San Francisco Administrative Code during May 1981 (Nelson Nygaard Consulting Associates, 2001).

Since the inception of the fee, transport operator Muni has collected $US93m (€104m). As of the financial year 1998/1999 the TIDF fund balance stood at $US56.5m (€63.5m). Interestingly, the TIDF has been challenged in the courts a number of times by developers, but so far has been defended successfully.

Elsewhere in the United States there are no similar schemes whereby the money is earmarked for transit. However, Redwood City in California did adopt an updated traffic impact fee to accommodate traffic generated by new development. While the majority of the $US12m (€13.5m) went towards road improvements, around a quarter was allocated to fund various transit and other transport demand management measures (Nelson Nygaard Consulting Associates 2001).

Developer charges in Toronto, Canada

In Toronto, Canada, development charges were put in place for projects on the Sheppard subway corridor and the North York City Centre on both commercial and residential projects, essentially because of the withdrawal of financing from the provincial government for capital projects. The charge was intended to recover part of the costs of the subway line and other infrastructure. Construction was started on the Sheppard line in 1997, and

Figure 3.6 San Francisco is one of the few cities where impact fees are collected to fund transit

was due to be completed in 2001. However, due to the deep recession of the early 1990s, and the effective opposition by residents to increased densities, only a few developments on the corridor went ahead. Instead, the new, larger City of Toronto took advantage of new provincial legislation to apply the development charges across the city. Because of economic concerns, the levy is only applied to residential developments, and not commercial projects (Sims and Berry, 1999).

Developer charges through a Parking Place Directive, Hamburg, Germany[19]

Hamburg is a city of about 1.8 million people and covers 750 square kilometres. The Parking Place Directive provides both disincentives for car use and incentives for public transport use. It was enacted to try and ease congestion in central Hamburg by preventing new parking spaces from being developed. Previously, developers were required to provide parking spaces for residential and commercial developments. Now the developer is required not

Figure 3.7 The modern Hamburg Metro that receives funds from the impact fee based on parking spaces, which also help to create a better operating environment for public transport

to provide parking, but instead pays a special relief which is used by the city authorities to improve public and other more sustainable transport.

The directive was set up in 1992 and has operated, with some modifications since then. The legislative basis is the Hamburg Building Law 49, Compensation for Car and Bicycle Parking, passed on 15 April 1992. In the city centre (*c*. 8–10 square kilometres) instead of each car parking space, the developer has to pay €16500.[20] Instead of each bicycle parking space, the developer pays €1650. The money is paid to the city administration, which decides independently about how it is spent. In the last five years the fund has raised €51m.

The mechanism is seen as acceptable to politicians because everybody gains. The inhabitants have a better quality of life, visitors can walk around more easily, and the shop owners are selling more. To users, public transport in the centre is regarded as very good. No real problems were encountered, and compensation for parking provision is now part of the legislation of most German cities. In many cases it is the builder's decision whether he will build parking spaces or pay the compensation. The compensation costs less than building parking. In Hamburg the difference is that the developer is obliged to pay the compensation and cannot build the parking spaces.

Developer levies and charges: conclusions

Charges by local authorities as part of granting developers permission to build are common-place. Increasingly these are used to provide infrastructure support for public transport and, although it is more difficult, sometimes revenue subsidies as well. Developer charges are generally popular among voters because they are perceived as providing a benefit for nothing, as the developer (or the developer's client) pays instead.

However developer levies have a number of practical limitations. Useful though they may be for a particular locality, they are not available to fund public transport infrastructure and fares over a city as a whole. Furthermore they are not spread evenly across an area – meaning that improvements may not necessarily be made in the most deserving neighbourhoods. All this is well and good, but often it is in stable areas, in places where only limited development is taking place or in areas of decline, that there is the greatest need for public transport investment. There is also an issue of timing. Developer levies do not provide capital much in advance of development, unless planning permission is sold up-front, and it may be hard for localities to ascertain capital needs and thus the fees. Developer levies are also criticized for deterring development and increasing new housing costs, and resulting in competition between local authorities. Also, communities may change their policy preferences depending on economic conditions, for example, finding a need to subsidize new development rather than letting new development subsidize the existing community.

Overall developer levies are useful in selected cases but have severe limitations as a general source of public transport finance.

From a developer's perspective, impact fees may replace more unpredictable, negotiated exactions. On the other hand, UK experience suggests that developers are often more experienced at negotiating than local authorities, allowing them to reduce the amount they pay. One other benefit for developers of negotiated exactions is that local authorities in economically depressed areas are often so desperate for investment that they can be played off against one another, thus resulting in a more favourable outcome.

Exactions allow more flexibility than fixed impact fees. The problems are that exactions are not as predictable or equitable as developer charges or impact fees, due to their being individually negotiated. Fairness may be decreased if politics enter into private negotiations. In addition, the revenue source is only as predictable as the economic conditions affecting the amount of development.

Conclusions

All the LETs in this chapter seek, in some way, to capture a portion of the added value of public transport improvements from those who benefit from those improvements. There is a hierarchy of beneficiaries that form a system linking from the landowner of a site, its developer, the property owner, through to the business and people occupying properties, their employees, customers, users and visitors (Figure 3.8)

The first question is where within this system might a LET be introduced. If the LET is towards the top of the system, its cost could well be passed down, for example, by a landowner charging a developer more for a lease on a site, or a rise in the price a developer charges when selling to a property owner. The reverse is unlikely. If a LET is positioned at the top of the value chain its

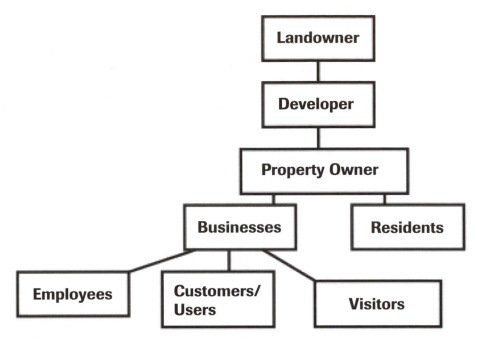

Figure 3.8 The system of local beneficiaries of public transport improvements

effect seems more likely to spread down through the system, with each level required to pay its share for the value enhancement of public transport.

In practice, various LETs are targeted all over the system, as is shown in Figure 3.9.

The distribution of the LETs is notable, and it is also notable that LETs have avoided some parts of the beneficiary system. LETs on individuals are rare (those on students being really minor exceptions on transient populations). LETs seem particularly concentrated in the middle part of the system – on developers and property owners. Why might this be? Is it simply politics, or is it practical considerations in designing a LET mechanism?

With rare exceptions, there tend to be few examples of any type of taxes on the value of land, and this lack of fiscal attention may help to explain why land value LETs are also rare. Yet, because land value is not subject to national taxation, this could present an opportunity for a local beneficiary pays LET in circumstances when public transport plays a role in enhancing land value.

LETs upon developers and the development process are widespread. Farrell (1999*b*) noted that, regarding the planning gain mechanism, the beneficial impact was highly localized and easily identifiable and there was a small number of players. All this made a LET easier to justify and not administratively complex. For land value it can be harder to make the case for a LET and, unless there is an administrative process in place, it can be costly and difficult to set up and run. Regulation of new developments occurs anyway, the planning authority is in a relatively strong position in such circumstances, and so it is a fairly simple process to add a LET on to this.

However, with a LET applied at only one point in time, and when, perhaps, the effect of transport improvements are not entirely clear, the yield can be limited and erratic. Farrell (1999b) concluded that such planning gain mechanisms can only provide a small part of the total budget for infrastructure for an area redevelopment. Developer LETs are not a substantial and reliable source of income, but can provide a useful supplement. Furthermore, where the transport investment is needed may not be where the planning gain occurs. For example, it may be necessary to invest in public transport in a city centre, but new developments are on the urban fringe, making available planning gain to address the adverse transport impacts there, but not to help the city centre.

Businesses offer an opportunity of a regular and possibly larger income source, as demonstrated by the variety of LETs on the number of employer taxes. Although these may be a tax on employees, it is the employer who usually pays. It is difficult to design a LET on individual employees other than as a local income tax (which is hard to apply in most countries). Employment taxes are also progressive but there is a danger that employment taxes could

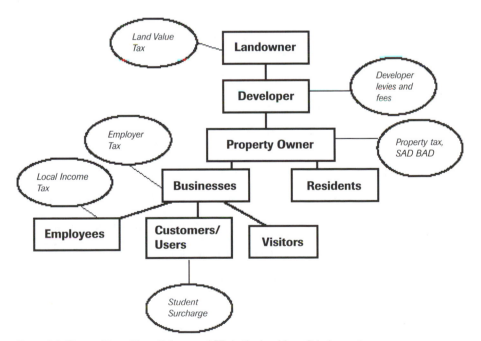

Figure 3.9 The position of beneficiary pay LETs in the local beneficiaries system

push firms to relocate to less public transport accessible areas to escape the tax. This raises a general point that applies to all LETs upon the lower part of the system – applied to employers, residents, customers and other users. If value to employers, residents and users outweighs the costs then all is well. If it does not, or there is a significant transfer of funds from losers to winners, then there is a danger of longer term structural readjustments. Employers may relocate their business; customers may go elsewhere. The projects that LETs fund must be efficiently designed and value for money produced. If LETs revenues are used to fund expensive and poorly patronized political prestige projects they could well fail. They may equally fail if they are used as a source of 'easy money' to avoid facing up to problems of inefficiency in public transport operations, poor management, restrictive practices, or cost escalations.

A positive LETs cycle could be generated, when efficiently invested LETs provide real benefits to businesses and residents, thus providing acceptability together with economic and social improvements to a town or city.

Notes

1   Based on Kramhöller (1999).
2   The figures have been converted from Austrian Schillings (ATS) to Euros at the rate of €1 to ATS13.8 (XE, 2001).
3   The figures have been converted from French Francs (FRF) to Euros at the rate of €1 to FRF6.5 (XE, 2001).
4   Based on Jones (1999), Rivenburg (1999) and Rice Center (1986).
5   The figures have been converted from United States Dollars ($US) to Euros at the rate of €1 to $US0.89 (XE, 2001).
6   Based on AC Transit (1999).
7   The figures have been converted from Canadian Dollars ($C-CAD) to € at the rate of €1 to $C1.4 (XE, 2001).
8   Based on Theile (1999) and www.bart.gov.
9   The following section is based on MTA for Los Angeles (2002).
10 Based on Tsukada and Kuranami (1994).
11 The figures have been converted from Japanese Yen (Y-JPY) to € at the rate of €1 to Y110.2 (XE, 2001).
12 The figures have been converted from Indian Rupees (RS-INR) to € at the rate of €1 to RS42.6 (XE, 2001).
13 The figures have been converted from Italian Lire (Lire-ITL) to € at the rate of €1 to Lire 1936 (XE, 2001).
14 The figures have been converted from Spanish Pesetas (ESP) to € at the rate of €1 to ESP166 (XE, 2001).
15 Based on Copenhagen Transport et al (1995), and Ahm (1999).
16 Based on Collins (2002).
17 The figures have been converted from British Pound Sterling (£) to Euros at the rate of €1 to £0.7.
18 Based on TCRP (1998).
19 Based on Gourd (1999).
20 The figures have been converted from Deutsch Marks (DM-DEM) to € at the rate of €1 to DM1.96 (XE, 2001).

Chapter 4

Polluter pays: Environmental policy emphasis

Pollution charges and practicalities

Chapter 2 contained a discussion of the 'polluter pays principle' (PPP) and the concept of a tax or charge to reflect the external health and environmental costs of pollution. By putting a price on polluting activities and technologies, pollution charges give producers or consumers incentives to change to cleaner alternatives. This improves the efficiency of the tax system, by accounting for the external costs of pollution, and can lead to savings over the cost of traditional regulation. Ideally a *Pollution Charge* should be based on the amount of pollution and other externalities generated. Strictly, a transport pollution charge should be on the amount of transport pollutants produced, for example on emissions such as sulphur, particulates, or carbon dioxide plus health costs, congestion, and all the externalities of transport discussed in Chapter 1. There are transport examples of pollution charge taxation measures, such as the UK's annual Vehicle Excise Duty, which is based upon CO_2 emissions. However it is often difficult to determine the true costs of pollution and thus the proper levels for pollution charges. For example, the UK Vehicle Excise Duty does not attempt to cost CO_2 emissions; it simply involves the redistribution of the tax yield of the old system. There are thus a whole range of practical issues that lead to approximations (rather than detailed cost-based data) being used to inform pollution charges (USEPA, 1999).

A practical issue is that it is often administratively easier to charge not pollution, but pollution-generating activities or other proxy indicators. Thus a tax or charge may be upon distances driven in a car rather than the pollutants emitted. In practice 'polluter pays' LETs tend not to charge on pollutants, but on polluting activities, although they may incorporate adjustments (such as alternative fuel vehicles attracting exemptions). This Chapter contains examples of LETs whereby transport polluters pay local earmarked charges that aim to curb externalities of car use such as emissions and congestion, and where some of the funds generated are used to finance public transport.

'Polluter pays' LETs appear to be the sort of fiscal instruments that fit most closely with modern transport and environmental policy principles. They have attracted considerable attention in recent years with proposals to introduce measures such as road user charging. But, it should not be forgotten that direct charges for road use are very old. Toll roads and bridges have existed for hundreds of years and millions of motorists are charged daily for parking. However, these have been based upon the 'beneficiary pays' public good concept whereby road users pay for the road and parking services they use. They are nothing to do with the polluter pays principle. Indeed, most of the LETs considered in this Chapter are not polluter pays mechanisms. However, one important issue raised here is that it is not only possible to introduce new LETs, but to undertake an eco-taxation reform of existing charges for road use and parking. A number of 'old' LETs designed simply to charge road users appear to be evolving in response to the modern transport policy agenda.

Given that the 'ideal' pollution charge, directly upon the externalities of transport, does not exist, the LETs in this Chapter will be examined for how they are changing to:

◆   charge directly pollution, congestion and other externalities;
◆   charge by pollution-generating activities (e.g. distance driven, entering congested areas);
◆   favour 'greener' travel behaviour;
◆   manage travel demand.

The first mechanism examined in this Chapter is the use of revenues from parking. Parking charges (and fines) are a normal fact of everyday life but are only seldom used to fund public transport, or overtly to manage transport demand. However we have identified a number of examples of where parking charges and fines have developed to become a fully-fledged LETs mechanism.

The second group of LETs involve road user charges. This includes a number of new LETs that have been developed to address modern transport and environmental policy needs. Road users pay a charge for using road space in the form of annual registration taxes together with fuel taxation. These forms of taxation are not what economists generally understand by the term 'road pricing' or 'congestion charging'. In particular, the charges levied on road users relate very little to the costs of providing and maintaining the infrastructure, let alone to wider notions of optimizing its use either from a purely traffic perspective or from a much wider social and environmental perspective. Polluter pays road user charges involve more variable and targeted taxes. Congestion pricing and tolling are examples which have been implemented, although only on a limited scale. These schemes benefit public transport indirectly (by raising the cost of car use relative to public transport) and many are part of a package including revenue use for public transport. Several schemes will be discussed, including both successes and, importantly, some failed schemes.

The third group of LETs relate to general motoring taxes. Motorists are a large source of general taxation revenues throughout the world. In particular fuel is taxed, and in most countries vehicles are subject to license fees. Such taxation is usually levied at the national level and there is no hypothecation (earmarking) to a particular use. There are, however, some exceptions to this, particularly in the United States, where powers to levy local motoring taxes have led to some examples of where revenue has been earmarked to fund public transport.

Finally there is an example of where revenues from landing fees at airports are used for public transport. The environmental impacts of air transport are a growing concern and one example was identified of where airport charges were used to improve public transport to an airport. However, this example does raise some important issues about whether this sort of LET can be justified from an environmental perspective.

Parking charges, levies and fines to fund public transport

Parking charges are a normal fact of life and are used by local authorities and businesses as an income flow to fund their activities. Simple, cheap and quick to introduce and operate, parking charges are readily understood and accepted by the public. They can also be said to conform to the 'polluter pays' principle of taxation. Although parking is only a 'proxy' indicator of pollution, parking charges have for long been used in many cities to control the level of traffic and to discourage the use of cars for commuting and some other purposes. Importantly for a LETs mechanism, parking charges can provide a steady and continuous flow of money. Typically, though, parking charges are not widely hypothecated to support local public transport or as part of a planned transport funding package.

This section will explore a number of cases where money raised from parking charges has been dedicated to fund public transport improvements.

There are three different forms of parking payments that raise revenues for public transport:

- parking charges
- parking levies
- parking fines

Parking charges are the revenues from on-street and off-street parking that is open to all members of the public, whereas parking levies largely concern private parking, at workplaces for example. Parking fines may formally involve the legal system to punish parking abuse, but in practice often act more like an additional charge.

Parking charges at Heathrow, Gatwick and Stansted Airports, London, UK[1]

At the British Airports Authority (BAA)-run airports of Heathrow, Gatwick and Stansted, a proportion of the parking charges pay for improvements to public transport. It is notable that this is part of surface access strategies for these three airports that has been implemented by their private sector owners, and is not a scheme initiated or run by local authorities or the public sector.

At these airports, an average of £0.25 (€0.35) for every passenger's car parking transaction is earmarked to pay for improved public transport. This varies between £0.20 and £0.40 (€0.27 and €0.55) per transaction in short-stay and £0.30 (€0.41) in long-stay parks. This is credited to a BAA budget that goes towards improving public transport within and around each specific airport. In addition, £10 (€14) of the annual staff car parking pass at Heathrow, Gatwick and Stansted, is earmarked to improve public transport access.

The parking concessionaires (firms such as Pink Elephant and National Parking Corporation) collect the money, take their administration fee, pay Value Added Tax,[2] and then hand the balance to BAA, which transfers it from its parking budget to a public transport improvement budget. This is administered by the transport managers at the three airports, and goes to

Figure 4.1 Car parking at Stansted Airport. Part of each parking fee is earmarked to improve public transport access

promote public transport networks through improved marketing, and where appropriate through improved physical measures (e.g. bus lanes, signal priority for buses etc.).

The idea for introducing a dedicated parking levy first arose at BAA during 1995, while a national debate on motorway tolls was underway. Following this, the average £0.25 (€0.35) charge on passenger parking was introduced at Heathrow in April 1996, with Gatwick following in June 1998 and Stansted in July 1999. The staff levy was introduced at Heathrow, Gatwick and Stansted in 1999. Interestingly, most employers on the airport sites do not pass any of the cost of car park passes on to their staff. They thus have no transport effect upon the travel behaviour of individuals. The 'polluter pays' effect on staff at the airports is therefore all but eliminated.

Unsurprisingly, political acceptability of the mechanism is very high among councils and operators, although within BAA there was a difficulty in justifying why it, as an airport operator, was charging customers to fund public transport. However, the rationale for this is now accepted. On the user side, no complaints were encountered, probably because only a very small amount is involved – and in many cases users are simply unaware of the additional charge or its use. The operation of the mechanism has also proved relatively trouble free.

Overall, the approximate revenue raised from the levy for 1999 at Heathrow was £2m (€2.8m), out of a total public transport expenditure of £650m (€930m). Public transport expenditure at Heathrow in 1999 was exceptionally high, as the Heathrow Express rail project was just opening (see below). But even allowing for this, the parking levy raises only a very small proportion of the funds needed for public transport development. Revenue raised at Gatwick is in the order of £1m (€1.4m) per annum, and that at Stansted is in the order of £250,000 (€357,000).

It is important to note that the LET funding from the parking charge to the general public and the levy on staff parking is part of a broad range of surface access measures. At Heathrow, complementary measures to improve public transport access to the site have included the construction of the Heathrow Express Link (to Paddington), and a network of bus lanes across west London. An extension of both the Heathrow Express and the Piccadilly Line to serve the new Terminal 5 is also planned. In addition, the Central Bus Station has been modernized and tougher parking policies and improved traffic control measures have been introduced on site. The transport fund also paid for the UK's first motorway bus and taxi lane on the Heathrow M4 spur road.

At all three airports, staff have been encouraged to use public transport through the introduction of a new *Airports Travelcard*, providing subsidized fares. BAA Stansted has also adopted a policy whereby airport staff recruitment takes place on public transport corridors. For instance, local towns and

Figure 4.2 The Central
Bus Station at Heathrow
Airport. Revenue from car
parking fees helped fund
its modernization

villages to Stansted Airport now have such high levels of employment, that
the airport has been forced to look further afield. Accordingly, areas along
the rail line into London, such as Harlow, and Tottenham Hale in north-east
London (where unemployment is far higher) are being studied.

The small level of the charge, the fact that it contributes only a very small
proportion of the budget spent by BAA on public transport, and the fact
that many employers do not pass the charge onto car commuters raises the
question as to whether this really is a 'Polluter Pays' mechanism. It could be
viewed more as a 'Spreading the Burden' source that just happens to have a
small polluter pays effect.

Parking charge in Aspen, Colorado, United States[3]

The case of Aspen, Colorado shows that even in the car-centric United States,
parking levies are seen as an acceptable way of raising money, although,
initially at least, a great deal of opposition had to be overcome. Parking policy

in Aspen is intended to address a series of objectives. These are to increase the proportion of spaces used by shoppers, decrease congestion, and improve the environment, by holding the parking supply steady and using the levy to discourage the number of single occupancy vehicles. In the town centre, motorists may only stay for a maximum of two hours, and pay $US1 (€1.1) an hour to park. Outside the centre, in the residential district, non-permit holders pay $US5 (€5.6) a day. The parking meters use smart card as opposed to credit card technology, meaning that the money is transferred immediately, and administration costs are lower.

Parking revenues are paid into an Enterprise Fund, from which money is earmarked to pay for transport alternatives. The fund pays for the marketing of transport alternatives, for free buses, and contributes towards a $US85m (€95m) 71 km light rail system. The light rail will connect with the major Colorado Interstate highway and Amtrak rail system, via five park-and-ride sites. The light rail scheme is awaiting Federal approval. Parking revenues generate about $US1.6m (€1.8m) a year, of which $US600,000 (€672,000) is put aside for the proposed light rail system.

The existence of the fund means the city's Transit Department does not have to fight for its money every year. Another key benefit relates to the unique US system requiring local referenda for expenditure plans. Now it is set up and approved, there is no need for voter approval every time a new idea is tried out, just a decision from the City Council. This mechanism does incur some additional costs to the municipality, such as paying the finance department for processing the money and the police department for time spent on traffic duties to enforce the parking charges.

Parking charges are still very controversial in the United States. On the day before the scheme's introduction, car users in Aspen held a 'honk-in' at noon, the noise being reported as 'deafening.' In addition, a cardboard effigy of a parking meter was burnt. Despite this, five months later, voters decided by 3:1 margin to keep the scheme. Parking spaces, which previously had an occupancy rate of 98% fell to 63%, even in the peak. This means that spaces are not being tied up all day by commuters, and there are more places for shoppers to park. This is an important consideration for businesses where many shoppers spend around $US2,000–$US3,000 (€2,200–€3,300) in an afternoon! Previously, there had been very few parking spaces available for shoppers, as 70% of downtown spaces were occupied by commuter vehicles.

Since the introduction of the parking charges and the various measures, such as the free bus system, bus ridership has increased by 35% and 4.5m trips are now made on the system each year. The impact of free bus service was enhanced as the previous way of charging fares was awkward and inflexible. As is common across North American public transport systems, there was a flat fare for a fixed time period. Passengers had 90 minutes after paying a fare to complete their journey, before they had to purchase another ticket. This was too short for shoppers, and also discouraged commuters from using buses.

Parking charges in La Spezia, Liguria, Italy[4]

A parking charge dedicated to public transport was also adopted in the Italian city of La Spezia. This was introduced following two significant shifts in legislation which transferred responsibility for funding local transport from the national government to regions and/or local authorities.

In order to address their new legal responsibilities, in particular the problems of pollution and the lack of parking in the City, the administrators of La Spezia, aimed to encourage commuters travelling to the city centre to transfer from the car to public transport. The scheme involved the regulation of parking areas with the intention of making public transport competitive with private transport. The new parking plan came into effect at the beginning of February 1999. It provides toll parking for about 10,000 cars (which had previously parked free of charge) in the centre of the city, with decreasing rates toward the suburbs. Special exemptions apply for residents.

As is required by the enabling laws, the first call on the parking income is to enhance parking. Two-thirds of the money collected (of a total estimated to be €770,000) is being used to provide new parking areas, while the rest goes to finance a free bus service connecting the centre of the City with two park-and-ride areas where users may park for free. Each free parking area accommodates more than two hundred cars. The two bus lines transport about eight hundred passengers each day, and the two parking areas are almost full during the whole day. The majority of the users of the free parking and bus services are commuters. The free bus service in La Spezia is organized by the ATC operating agency, which also administers the bus service and toll parking in the centre of the city.

The La Spezia examples are not the first of a parking charge LET in Italy. In the City of Verona, the local bus operator (AMT) has, since December 1992, administered a 260-space toll parking area located not far from the centre of the city. To enter the AMT parking area, the users have to buy a daily bus ticket (it is not possible to buy a season bus ticket to enter the area). Of course in the centre of the city, there are parking areas other than the one administered by AMT, so this is simply a one-off scheme operated by the owner (AMT) who have no plans to operate any further parking areas.

'Mobilityfund', Amsterdam, The Netherlands

A Dutch example of using parking charges is in place in Amsterdam. Here a light rail line (the Ijtram) from the Yburg housing estate to the city centre is to be financed from a 'Mobilityfund', the majority of which is derived from parking charges. For example, in 1999 parking revenues amounted to €13.1m of a total fund of €18.6m (Gemeente Amsterdam, 1999).[5] Of this, €341,000 is to help pay for the tram line.

Overall, Gemeente Amsterdam, the municipality, will contribute only 5% of the capital cost, with the rest financed by the national government. The total cost for the Ijtram is estimated at around €90.6m. It was decided that Amsterdam should contribute in financing this tram. An amount of €1.7m will be deducted from the Mobilityfund over five years starting from 1998.

Additional parking charge examples

Revenues from city-centre parking and suburban park-and-ride schemes are also used to fund rail infrastructure in Milan (Farrell, 1999a), while public transport benefits financially from parking revenues in Jacksonville, Florida, and San Francisco (Bushell, 1994). In Germany, amendments were made to the German Road Traffic Act in August 1994, which made it possible for local authorities to use the earnings from parking spaces to finance public transport infrastructure. Previously, charges had to be used only to improve parking facilities (Copenhagen Transport et al., 1995).

In the Greater Vancouver Region, the estimated revenue from a provincial sales tax on commercial parking is $C7.5m (€5.3m). This revenue is transferred from the Province to the Greater Vancouver Transportation Authority (GVTA), under the GVTA Act.

An interesting example of the use of parking charges comes from Milton Keynes in the UK. Until recently, Central Milton Keynes had free parking provided for all employees and customers of the businesses locating there. The parking spaces are owned by the local authority (Milton Keynes Council) who started to introduce charges to a small proportion of spaces. However, local businesses were unhappy about these charges. Under the Milton Keynes Economic Partnership (MKEP) the parties involved came to an agreement to maintain the competitive edge of Central Milton Keynes while meeting the overall transport strategy for Milton Keynes. From this has developed the Central Milton Keynes Transport and Parking Strategy, a plan for the period up to 2011, including modal shift goals, which include cutting car-driver-only trips from 70% now to 50% and raising bus use from 9% to 20%. The Strategy involves a mixture of incentives and disincentives to achieve this, including the parking charges paying for the measures such as a Park and Ride bus service and an area Travel Plan to reduce car commuting by employees. The latter includes, for example, an inter-firm car sharing scheme. An innovative feature is that the finances for this Strategy are channelled through a charitable Partnership Company.

Parking charges: lessons

Parking charges are commonplace. These examples demonstrate how in some cities, they are
being reformed into LETs mechanisms and integrated into local environmental and transport
policies. Parking charges are a proxy 'polluter pays' mechanism that can be a very effective
tool of demand management. However they seem to only raise a limited (though often useful)
amount of revenue to support public transport.

Parking fines

If motorists evade a parking charge they may well end up paying a parking
fine. Parking fines may formally involve the legal system to punish parking
abuse, but are sometimes technically a higher charge. In some places, revenue
from parking fines as well as parking charges are earmarked to fund public
transport.

In France, additional revenues from parking fines and driving offences
have been earmarked to pay for public transport infrastructure since 1973
(Ministère de l'Aménagement du Territoire, de l'Equipement et des Transports,
1995). This was enabled by the passing of the same piece of finance legislation
that resulted in the *Versement Transport* in 1971 (see Chapter 3) (Meyer,
1996). The money is paid directly to the Communes and their associations
when their population is above a certain level (currently 10,000 people), and
indirectly on a proportional basis when the population is below this level.

In the special case of the Ile-de-France region, 50% of the money is
allocated to the Syndicat des Transport Parisens (STP), 25% to the Region,
and the rest to the local authorities. The STP element from the fines is
earmarked to finance projects which improve either a connection between the
different transport modes (interchanges, regional car parks, etc.) or operation
of transport networks, and accessibility to the network (passenger transfer
tunnels, travelators, road crossing improvement, etc.). These subsidies
are matched by the Ile-de-France region (Ministère de l'Aménagement du
Territoire, de l'Equipement et des Transports, 1995).

In Athens, Greece charges imposed on private cars that violate bus lanes are
passed to OASA – the public transport authority for the Athens Metropolitan
Area (Patrikalakis, 1999; Mitoula *et al.*, 2003).

In the UK, any excess money resulting from fines collected in areas
designated as either a Special Parking Area (SPA) or a Permitted Parking
Area (PPA) – whereby parking offences are decriminalized and become the
responsibility of local authorities – are retained by the highway authority.
This must then be specifically used to provide parking facilities, build road
improvements, or enhance public transport. SPAs and PPAs were enabled in
the 1991 Road Traffic Act. The first such area was in Wandsworth in late

Figure 4.3 Another parking fine . . . but in some locations part of the fine supports public transport

1993, the rest of London having followed suit by July 1994. As of June 2002, eighty-five local authorities, including Winchester, Oxford, High Wycombe, Maidstone, Watford, Luton, Portsmouth, Manchester and Edinburgh have decriminalized parking enforcement powers, including the thirty-three in London (*Parking Review*, 2002).

Parking fines: lessons

Like parking charges, in some cities, parking fines are being reformed into LETs mechanisms and integrated into local environmental and transport policies. Their impact is probably stronger as a fund raising mechanism than as an influence on behaviour.

Parking levies

Controlling access to parking spaces has long been considered one of the most effective tools at the disposal of local authorities to reduce car use. But, councils have been hampered because typically the majority of parking spaces in town and city centres are privately-owned, making it extremely difficult for any parking control policies to be implemented. In Britain, the workplace parking levy, which provides local authorities with optional powers to charge employers a levy according to the number of employee parking spaces, was first proposed by government in the consultation document 'Breaking the

Logjam' in 1998 along with road user charging (DETR, 1998*b*). In England and Wales, the levy became law through the 2000 Transport Act, but in Scotland the workplace parking levy option was abandoned, and is absent from the parallel Transport Act in Scotland. As yet no local authority in the UK has introduced a workplace parking levy, although Nottingham and Milton Keynes are among several who are actively considering this measure. It is of note that Paris is considering the introduction of a workplace parking levy to supplement income from the *Versement Transport* employee tax. This is because more money is now needed than the *Versement Transport* can provide.

Parking licence fee, Perth, Australia[6]

Perth, in Western Australia, is one of the most car dependent cities in the world. In 1996, 91% of households in the State had at least one registered motor vehicle. In 1991 the State Government of Western Australia, in partnership with the City of Perth, first identified the workplace parking levy or 'parking licence fee' as one of a number of measures that could ensure a better set of access and amenity outcomes. However, it was not until 1996 that the licence fee scheme obtained State cabinet approval as a component of the wider Perth Parking Policy. This in turn forms part of the Perth Metropolitan Transport Strategy. It then took until 1997 for the licence fee scheme to be adopted by the City of Perth. Legislation to provide the basis for the Perth Parking Licence Scheme entered the State Parliament in November 1998 and became law in July 1999.

Within the Perth Parking Management Area, all parking, both on-street and off-street, except private off-street residential is licensed. Thus the parking licence fee is a private-non-residential parking levy (rather than only applying to the workplace as is the case for the UK Workplace Parking Levy). Although all parking is licensed, fees are not charged where parking spaces:

◆   help the city work – e.g. loading/unloading spaces;
◆   promote access – e.g. bus layovers;
◆   provide a community service - meals on wheels, patient transfer services, blood transfusion services; or
◆   an incidental to the prime business activity – e.g. car sales and service.

In addition, small businesses with less than six parking licence fee liable bays on their property were required to licence their parking, but exempted from the parking licence fee. In total, these exemptions applied to around 6,000 of the 58,500 licensed spaces, of which 4,000 were exempted on usage grounds, and the remainder due to the 'small business' rule. From an administrative perspective the small business rule reduced the number of licence holders liable to pay the fee by more than a third, for a relatively small reduction in revenue. The justifications for all spaces being licensed even if not liable for the parking licence fee, were so that the scheme, and parking in general, could be monitored and enforced more effectively.

Government bodies are not exempted from the charge, and must be licensed and pay the same licence fee. This means that the largest single payer of the licence fee is the City of Perth, controlling as it does two-thirds of all public off street parking and all the on-street parking. Overall, of the 58,500 licensed spaces, around 6,000 are on street and the remainder off-street. Nineteen thousand off-street spaces are public, while the remainder are tenant parking spaces not available to the public.

Legally the licence fee is a tax, for which property owners rather than tenants are liable, because they are less mobile, easier to trace, and there are fewer of them. In practice, it is the tenant who pays the parking licence fee, and it is common practice to have a clause in a tenancy agreement that the owner can pass on any government charges or taxes. Thus the user of the parking space normally has to pay the fee. This stands in contrast to, for example, the Heathrow parking fee mechanism where employers pay the fee and so the effect on end users is nullified. Under the Act, the rate per space was set at $A70 (€35) per year when introduced in 1999.[7] This was increased for the 2001/02 licence year to $A120 (€60) a year (*pro rata*). To enforce the parking licence fee, the legislation allows authorized inspectors to enter property and demand records.

In Perth, the money raised must be spent improving the access and amenity of that area, and as a result it is earmarked to fund the Central Area Transit (CAT) bus system. It is believed that this clear link between charge and benefit is why the expected opposition to the fee did not really materialize. Around 80,000 people a week use the two state-of-the-art CAT services. Perhaps also of significance in the acceptability of the parking licence fee in such a car dependent city, was the existence of the Free Transit Zone (FTZ), which was originally established in 1989. In conjunction with the introduction of the parking licence fee, this was expanded to cover the 825-hectare Perth Parking Management Area. Altogether, 45,000 people use buses and 15,000 use trains in the FTZ each week.

Around 56,300 spaces were licensed during the first year of operation, generating $A3.35m (€1.67m). Non-payment at $A65,000 (€32,500), was less than 2% of the total due. One impact of the scheme was that parking supply fell by nearly 10%. There are 6,000 fewer spaces than recorded in a 1998 parking survey. Most of the spaces taken out of use were situated near the edge of the Parking Area and remote from areas of high parking demand. There is also evidence that small businesses were decommissioning spaces to meet the five spaces or less exemption, and that property owners are far more likely to act to stop people illegally using their spaces.

Parking space levy, Sydney, Australia[8]

Although possibly less well-known than the Perth example, the so-called Parking Space Levy (PSL) in Sydney was actually introduced several years earlier, beginning operation in the Sydney central business district and North Sydney in July 1992. Under the Parking Space Levy Act of 1992, businesses were required to pay $A200 (€98) per parking space per year until July 1997, when this was increased to $A400 (€195) per space per year – much higher than in Perth. In May 2000, the Parking Space Levy Amendment Bill 2000 increased the rate further to $A800 (€390). It also extended the levy to four other business districts in Sydney (Bondi Junction, Chatswood, Parramatta and St Leonards). A zonal system was also introduced, with these new business districts being referred to as Category 2 areas, with a lower levy per space per year of $A400 (€195). Sydney CBD and North Sydney are now referred to as Category 1 areas. However, there are several categories of spaces that are exempt from the charge. These include spaces designated for registered disabled people, residents, charities; or for loading/unloading bays. Parking Space Levy fees are collected by the NSW Office of State Revenue (OSR) on behalf of the NSW Department of Transport. Any business within one of the six designated PSL areas must register with OSR, and make PSL payments to OSR on the basis of their liability.

Unlike in Perth, the PSL applies only to off-street private parking used by tenants of commercial office buildings, and requires the owner to pay a tax on all parking spaces on their property regardless of whether they are used or not. A further important point is that all public car parking is exempt. It could also be argued that Sydney is more orientated to raising revenue (although it would have been hard to raise less revenue than in Perth). Interestingly, the terms of the Act allow the revenue only to be spent on infrastructure and maintenance, and not on subsidizing operations. Although this is seen as being restrictive, there is also a balancing view that this provision does help prevent the levy being used to replace public transport funding from general funding sources. As a result, the funds raised from the charge have been spent on improving:

◆   interchanges – Bus/Rail, Bus/Ferry, etc. at locations that serve the levy areas;
◆   car parks within areas from which commuters travel to PSL areas – but outside PSL areas;
◆   public transport infrastructure, such as the development of Rapid Bus-

Only Transitway bus stations and light rail, that provide services within
or to/from PSL areas; and

◆   electronic passenger information systems for Transitway interchanges.

The revenues collected since the introduction of the Levy in 1992 have
grown mainly from the increases in the levy charge per space, the extension of
the scheme in 2000, and from new development within the PSL areas. Money
raised in the 2000–2001 financial year was roughly $A40m (€20m). One
potential problem concerned the boundary locations. In the event, council
zoning boundaries were used and this appears to have been successful. Other
issues have centred on how much to charge – spend too little and car use
will not be affected, but spend too much and businesses think of moving or
closing – and exemptions. In particular, there is disquiet that retail car parks
in Category 1 areas are not exempt, whereas in Category 2 they are. So far, no
action has been taken, as the State Government regards this as a characteristic
of the two areas.

Commercial parking tax, Vancouver, British Columbia, Canada

One other slightly different tax that is dedicated to transit, is the commercial
parking tax in Vancouver in Canada. This is a provincial social service

Figure 4.4 Buses in Central Sydney, Australia. Bus/Rail interchanges have been part funded using
the Parking Space Levy

Figure 4.5 A bus in Seattle, Washington State. Part of the U-Pass scheme

tax, introduced by the Social Service Tax Amendment Act 1993, that was originally levied at a rate of 7% from June of the same year on the purchase of motor vehicle parking within the Vancouver Regional Transit Service area (Ministry of Provincial Revenue, 1993).

U-Pass scheme, University of Washington, Seattle, Washington, USA

In Washington State, one programme part funded by parking income is the U-Pass. This is a flexible transport benefits package that offers University of Washington students and staff a variety of commuting options at a greatly reduced price. Introduced by the University of Washington and the Municipality of Metropolitan Seattle in 1991, the pass is part-funded by revenue from parking fines and from a parking levy. The programme was developed in response to concerns for trip reduction and improved commuter services in view of possible impacts from planned campus development. Overall, the annual budget for U-Pass in the 1999–2000 financial year was for $US9.9m (€11.1m), of which 88% paid for the fifty bus routes that serve the campus. Of this, $US462,000 (€517,000) was from dedicated parking fines and a further $US4.1m (€4.6m) was derived from a hypothecated parking levy (University of Washington Transportation Office, 2000).

Parking levies: lessons

Like parking charges, parking levies are a proxy 'polluter pays' mechanism. However, they are largely 'new' LETs, having been developed in response to modern transport policy needs. The success of parking levy LETs in very car dependent cultures suggests that this is a mechanism that could be widely transferable. But it is also a mechanism that requires great care in its design and implementation. Both Perth and Seattle show the relation between the introduction of the parking licence fee and the improvement to public transport funded. Any area parking levy system needs to be carefully justified, with targeted exemptions to cover equity issues – both for social reasons and for any major 'losers'. In design, it should be as simple as possible to understand.

The three examples of Perth, Sydney and Seattle show great flexibility in how a parking levy LET can be designed. In Perth the parking licence fee is very low but spread over a broad base of payers. In Sydney the charges are much higher from a narrower base. In Seattle the LET was at the level of a university site. The Sydney levy raises a significant sum of money, as did (at its own level) the Seattle scheme, whereas Perth's income is small. Yet, despite the low level of the charge, the impact on the amount of parking in Perth was significant (although only initially).

Although classified as polluter pays schemes, there are 'beneficiary pays' elements in these cases. The public transport improvements funded by the charges clearly needed to be of the order to win acceptance from business and users. As was noted in Chapter 3, proceeds from the levies need to be seen to outweigh the charges paid. The very low charges in Perth certainly achieved this – indeed there seems to be a strong 'spreading the burden' element in Perth, as the charges are nowhere near high enough to fund the public transport improvements and free buses. In Washington the scheme was clearly one where user benefits were closely linked to its design.

Charging for the use of road space

As noted at the beginning of this Chapter, the idea of charging for the use of roads is far from new. Already in the late seventeenth and early eighteenth centuries many roads were built as private toll roads. However these were essentially user charges, much like fares on ships, trains and public transport. Tolls have been charged on roads, tunnels and bridges to offset the expenses of new construction, operation, and maintenance and are also imposed on boats (e.g., docking fees) and aeroplanes (e.g., landing fees).

In general tolls have not been used to charge for pollution, to favour 'greener' travel and certainly not to manage demand (indeed, the greater the use of a road or a bridge the better as it increases revenue). But tolls, and other charges for roadspace, represent an important area of LET development. In some places, traditional road charging methods, like tolls, have evolved to become a transport policy LET, and new methods of road charging have been developed specifically as an instrument of transport demand management.

There are different forms of charging for the use of roads. The terms *road taxes, road pricing* and *congestion pricing* are often confused. *Road taxes* are levied on road users, some of which relate to the extent of their road use, like fuel taxes, and others that may not, like licence fees. *Road pricing* is commonly used when particular road trips are subjected to well-defined charges, like road tolls or tolls for a bridge or tunnel. *Congestion pricing* is a particular form of road pricing that imposes higher charges on motorists who travel at times and places where the road is congested. An example may be to enter a city centre during the day. From a theoretical point of view these are efficient tools to let the polluter pay. All these forms have the potential to raise large sums of money of which parts are occasionally dedicated to support public transport.

Most often road space charging schemes were implemented as a package whereby the road pricing goes together with supporting measures for public transport. However a notable exception is Singapore's road pricing scheme – probably the best-known such scheme of them all. This began as an Area Licensing Scheme in 1975 as a demand management measure before being modified into an Electronic Road Pricing Scheme in 1998. It has succeeded in cutting road traffic in the city centre and holding it at that lower level for over a quarter of a century. No other traffic management system in the world has achieved anything like this performance. However, the funds from the charge are general revenue to the Singapore government. There is no earmarking to public transport or any other use. The Singapore road user charging scheme is not a LET (Wong *et al.*, 2002).

Figure 4.6 An Electronic Road Pricing entry gate in Singapore. These are purposefully conspicuous to ensure that motorists know they are entering the charge area

However, there have been LETs based upon Singapore's success. This section first considers the urban cordon road tolls in Bergen, Oslo and Trondheim in Norway where revenues are hypothecated to fund public transport, and in the UK. As in Singapore, tolling is based here on a cordon system, in which vehicles must pay for entry to the city centre, and the revenues are intended to fund a mixture of road and public transport investments, including safety and environmental improvements. There are also schemes that failed to be implemented.

These cordon pricing schemes are followed by a discussion of more conventional road tolls, where road users have to pay according to their use of a specific road, bridge or tunnel. The high occupancy toll lane (HOT-lane) near San Diego in the United States is one example, as is the Golden Gate Bridge toll in San Francisco where bridge tolls are used to subsidize inter-county traffic services, including bus and ferry. Given that the peninsula on which San Francisco stands is linked on all but one side by toll bridges, this could almost be viewed as a road pricing cordon in all but name. A wide variety of further case studies exists in the United States where toll revenue from bridges or tunnels are used to finance public transport. The Golden Gate Bridge example gives a good overview of the structure and mechanisms used also in other States.

Figure 4.7 The charge display board for Electronic Road Pricing in Singapore

The Oslo road toll scheme, Norway[9]

Oslo's road tolling scheme began in February 1990 as a major revenue generating element of a large package of transport improvements emphasizing new road capacity, safety and environmental improvements, and public transport. This was the second of three city-wide tolling schemes (Bergen began operating in 1986 and Trondheim followed Oslo in 1991), the Oslo scheme represents the first European attempt to charge a cordon toll for a large metropolitan area and the first implementation anywhere of electronic pricing on a large scale.

Due to the geographical layout of the Oslo, there are only three traffic corridors into the city. This meant that only nineteen toll stations and four street closures were needed to control virtually all traffic crossing a cordon line surrounding the central city. Toll collection and financial operations are carried out by A/S Fjellinjen, a corporation owned jointly by Oslo city and Akershus County. Users have three options for payment: manual collection by an attendant, payment to a coin machine, or electronic payment. Motorists opting for electronic payment can either purchase a seasonal pass or pay per trip at a substantial discount.

Waerstad (2002) views the successful implementation of the scheme as depending on four key factors. Firstly, the road tolls were introduced 14

Figure 4.8 A Modern Tram in Central Oslo. A proportion of the road tolls help support public transport, but most of the funds are devoted to roadbuilding

days after the opening of a major road improvement, the Oslo Tunnel (now the Festningstunnelen or Castle Tunnel). This was the main part of the 'Oslo Package I' – an integrated group of transport improvements which required the road tolls as part of their finance. Secondly, the State agreed to match fund the revenues raised by the tolls. Third, Norway has a 60-year tradition of roads being paid for by road tolls. The Oslo cordon scheme, although unusual, was not that different from the way motorists were used to paying for road access. Finally, the strongest opponents of the road building element were placated by the promise of 20% of any revenues raised going to pay for improved public transport infrastructure. Specifically, this has been spent on metro lines, metro stations, and public transport interchanges.

The Oslo Package II started November 2001 with an increase in the toll fee from NOK12 (€1.6) per passing to NOK12 (€1.9) while a levy of NOK0.75 (€0.1) was put on public transport journeys.[10] While the toll ring was supposed to be withdrawn in 2007, this is now open to debate as the public seems to have largely accepted tolling. If tolling is abandoned, the Oslo Package II will be funded from public transport fares only.

Overall the revenue raised from the toll ring for Oslo Package I has increased from NOK813m (€108m) in 2000 to NOK830m (€111m) in 2001 to NOK1012m (€135m) in 2002, although it should be noted that VAT was put on toll collection services from June 2001. Revenue from the toll ring for Oslo Package II provided an additional NOK40m (€5.3m) in 2001, NOK150m (€20m) in 2002, and is estimated to be NOK200m (€27m) a year from 2003 to 2007. Revenue from the toll on public transport fares was NOK20m (€2.7m) in 2001, and should be NOK130m (€17m) until 2011 under the current plan. The costs of operating the toll ring amount to around 10% of the total revenues (i.e. NOK82m (€11m), NOK95m (€13m), and NOK102m (€14m) for 2000, 2001, 2002 respectively).

Road user charging in Britain

Under the 2000 Transport Act, local authorities may introduce congestion charging within the UK. The city of Durham was the first in the UK to implement a congestion charging scheme. Since October 2002 there has been a charge of £2 (€3) for motorists accessing the historic centre between 10am and 4pm Monday to Saturday. There is a single entrance/exit to the charging area that is controlled by an automatic telescopic bollard. The bollard is raised during the period of charging and drops when a payment is made or when a vehicle fitted with a payment transponder approaches. The road user charging scheme was introduced after extensive consultation with affected parties in the area and has enjoyed widespread support. The amount of traffic entering during the charging period has been cut by a remarkable

90% with no noticeable increase in traffic on roads outside the charging area (Ieromonachou, Enoch and Potter, 2003).

In early July 2001, London Mayor Ken Livingstone finally unveiled plans to charge motorists £5 (€7.10) to enter central London between 7am and 6.30pm under legislation enacted in the Greater London Authority Act 1999. This began in February 2003 and the charge was expected to raise around £130m (€185m) per annum earmarked to pay for improvements to the city's public transport system, although in the event this figure is likely to be rather less due to traffic levels being affected more than was forecast. As of mid-2003, traffic within the zone had been cut by around 20%, with reductions also noted in areas adjoining the charging zone. Technically the system is similar to that used in Norway, with enforcement by digital cameras on roads into the congestion charging zone reading vehicle registration plates to check that the appropriate fee has been paid. A wide group of vehicle users is exempt from the charge, including the disabled, emergency services, motorcyclists, key pubic sector workers, school buses and public transport. Registered local residents pay 10% of the fee. An enlargement of the Central London zone is under consideration and the introduction of a second congestion charging zone covering Heathrow Airport.

Failed road user charging schemes

There are also road user charging schemes that failed due to lack of political support.   Between 1983 and 1985, Hong Kong conducted an extensive

Figure 4.9 A warning sign of an entrance point to the London Congestion Charging zone

evaluation of cordon congestion pricing schemes (called Electronic Road Pricing, ERP) including the first large-scale field test of equipment to collect congestion tolls electronically. The proposal was ultimately abandoned mainly because of popular political opposition. An important lesson from this is the need to anticipate and resolve likely objections early in the planning process. Furthermore, it is important to have a tangible and credible plan for re-distributing the revenue to the public.

The so-called Dennis Package in Stockholm was a range of complementary transport policies put together to appeal across the political spectrum, and included plans to build a ring road and to implement a toll ring somewhat like those in Norway. Unlike in Norway though, the tolls would be higher, and there was an explicit goal of reducing traffic in the city centre. Revenues were to finance a broad package of investments but were not specifically earmarked to public transport, although improvements of public transport formed part of the package. In the end the package proved to be too complex and failed to

---

Congestion charging: lessons

Overall, there are important lessons to be learnt from the successes and failures of road user congestion charging schemes to date. These are not about the technology of road pricing, which has attracted much attention, but about how schemes are designed, the effective inclusion of user concerns and political sensitivity. Major factors that appear to be associated with success are:

- having clearly defined and complementary objectives;
- not trying to achieve too much in the early stages;
- achieving at least some of the benefits promised as quickly as possible;
- being supported by politicians of all persuasions;
- being seen to work properly and reliably;
- gaining the support of the public;
- be understood by the public;
- having flexibility to develop as circumstances, public attitudes, objectives and technology change, and of being tweaked to react to 'unexpected' events;
- offering realistic alternatives to travellers who wish to switch from driving into the cordon
- paying attention to details

London in particular illustrates the 'lack of an alternative' issue, where the Underground and mainline rail services are already operating at full capacity. In addition, the Mayor's 2001 Plan for London seems to focus on plans for increasing high profile (and expensive) rail projects, which are unlikely to come to fruition in under 10–15 years. The Mayor's policy of enforcing bus lanes, more local bus services and subsidizing fares will undoubtedly help, but will have a marginal impact on providing a viable alternative to car drivers commuting into central London. For example, no mention has yet been made of express buses relieving rail and Tube routes – a measure that could quickly provide capacity when road charging is introduced.

Fundamentally, public support is crucial to the success of this type of scheme. While users may be prepared to put up with technical glitches and various uncertainties of how the scheme works in the short term, some rapid improvements in the transport situation are needed if there is to be a 'long term' road user charging concept.

reach implementation stage. The Dennis package is now viewed as a missed opportunity for the development of an integrated urban transport plan.

Golden Gate Bridge tolls, San Francisco, United States

Based in San Francisco, the Golden Gate Bridge, Highway and Transportation District (GGBHTD) operates a ferry and bus service in the travel corridor between Marin and Sonoma Counties and is also the owner of the Golden Gate Bridge. Tolling was implemented upon the opening of the Golden Gate Bridge in 1937. This was to repay bonds raised for the bridge's construction. In 1969, when the original bonds for the Golden Gate Bridge were retired, the California State Legislature authorized the continued existence of the Golden Gate Bridge district.

A unified bus and ferry system, called Golden Gate Transit has been operated by GGBHTD since 1971. The buses and ferries have needed subsidies and because the Golden Gate Bridge District does not have the authority to levy taxes, it was decided to retain Bridge tolls and use surplus revenue to subsidize the District's bus and ferry services. In addition, further expenditure has proved to be needed on the bridge itself, which has reinforced the case for retaining and increasing tolls. In December 1990, financial projections showed that the District's Fiscal Year 1992 expenses would exceed revenues by approximately \$US14m (€15.7m). These projections were the first to include the newly planned seismic retrofit of the historic Golden Gate Bridge.

Figure 4.10 Toll booth on the Golden Gate Bridge, San Francisco

The existing toll system currently consists of eleven manually collected toll lanes located on the south end of the Golden Gate Bridge. No toll is charged on the two northbound lanes. Tolls are collected southbound, 24 hours per day, every day. The bus fleet of GGBHTD is exempted from toll, as well as certain area police vehicles and carpools (three or more people).

By 1999, a $US3 (€3.4) toll paid by drivers helped generate an income of $US58.5m (€55.5m), of which $US21.6m (€24.2m) was dedicated to transit, and $US7.4m (€8.3m) subsidized the ferry services (GGBHTD, 2000*a*, 2000*b*). By 2001, nearly half of bus and ferry operation is funded by Bridge tolls, with another 32% coming from transit fares, and the remainder being met by Federal and State subsidies.

Additional examples

This Golden Gate case study forms just one example of a wide variety of case studies in the United States on the use of toll revenues from bridges or tunnels to finance (public) transit. Other well-known examples are from New York, New Jersey (Delaware Rivers), and elsewhere in San Francisco. In New York, for example, surplus operating revenues from nine toll facilities (bridges and tunnels) are channelled to the local New York City Transit Authority, which operates regional commuter-rail service around New York City. Since 1968 when funds were first redirected to fund public transport, bridge and tunnel tolls have contributed $US5.6bn (€6.3bn) to subsidize fares and underwrite capital improvements (TBTA, 2002). In California there are various examples which can be mentioned, especially in the San Francisco Bay Area. One of these is the San Francisco-Oakland Bay Bridge. Here revenues were used to finance the construction of the BART regional rail system.

High occupancy toll lane (Interstate15), San Diego, California, United States[11]

Another slightly different Californian example, is the High Occupancy Toll (HOT) lane facility in San Diego, which originally opened in 1988 as a High Occupancy Vehicle lane on Interstate 15 for buses, vanpools and two-person carpools. In this form it was similar to many other restricted high occupancy lane schemes across the urban United States. What makes the San Diego example a LET is its evolution into a toll lane.

The two high occupancy lanes were built by Caltrans, the California Department of Transportation, and are unusual in that users are physically separated from the general purpose lanes by concrete barriers. Drivers can only enter and exit the lanes at a single point in each direction, with access controlled through a ramp metering system – which also gives priority to high occupancy vehicles. The HOT lanes currently run for 13 km in the median

Figure 4.11 The HOT lane on Interstate 15 to San Diego, segregated from other traffic (to the right)

of Interstate 15 to the north of San Diego, California, and are reversible. The lanes operate in the peak-flow direction, i.e. north to south in the morning and south to north in the evening.

Figure 4.12 HOT lane charge display

Figure 4.13 The Inland Breeze express bus part-funded by HOT lane revenues

To begin with, this was an ordinary High Occupancy Vehicle lane on an interstate highway. The proposal that single occupancy vehicles (SOVs) be allowed to enter the lanes on payment of a toll was first suggested in 1991, as only 50% of the two lanes' capacity was being used while adjacent general-purpose lanes were experiencing severe congestion during peak periods. A local councillor and member of the San Diego Metropolitan Transit Development Board, Jan Goldsmith, advocated the scheme both to fill up spare capacity and to generate money to be spent on improving public transport services along the highway and into San Diego. Thus, at its conception, this was far from being a traffic demand management scheme. Indeed, quite the reverse was intended! The HOT lane opened in December 1996

As drivers approach the HOT lane, variable message signs advise them of the toll to use the lanes. The level of this toll depends on how much spare capacity is available in the HOV lane, and varies from €US0.50 (€0.56) to $US4 (€4.5) in normal circumstances, with drivers paying more the busier the lanes. Around $US430,000 (€481,000) of the annual $US1.6m (€1.8m) toll revenue covers operating costs, and $US60,000 (€67,000) pays the California Highway Patrol to enforce the lanes. State law requires the remaining money to be spent on developing the express lanes and improving the public transport service along the corridor, specifically, the express bus service known as the Inland Breeze which began operating in November 1997.

Bridge and road charges: lessons

The example of Bridge Tolls in San Francisco (and also elsewhere in the USA) show the way in which a traditional road toll, implemented simply to finance the construction of a road bridge, has been developed into a modern LET. Indeed, the bridge tolls in San Francisco (and also in Manhattan) have, to a large extent, produced a cordon road charging scheme in all but name. There are limited access points to enter a city centre across a series of bridges. In effect, San Francisco and Manhattan has had a cordon road charge for longer than Singapore!

The Golden Gate bridge tolls started off in 1937, simply as a revenue stream to repay the loans for the capital cost of building the bridges. It was only because of the joint operation of bridges and public transport, and the need for subsidies for the latter, that the bridge tolls were retained to become a LET. At this point they were a 'spreading the burden' and 'beneficiary pays' LET. It was the only real option available, and was not coupled with any notion of traffic demand management. The bridge tolls were simply a good revenue source. It is only more recently that elements of the polluter pays principle have emerged, with adjustments to favour car pools and buses using the bridges.

Like many other US LETs, bridge tolls are a pragmatic LET that have, over the years, started to evolve towards being a transport policy instrument. Like many other LETS they are a regressive form of taxation (in that there is no relationship between the charge and ability to pay), but are progressive once the overall system of gathering and earmarked expenditure is taken into account.

Charging for the use of lanes, as opposed to whole roads, is not widespread, and yet appears to have potential for general application. The San Diego HOT lane has proved popular with users, while non-users are ambivalent. One reason is that drivers are offered a genuine and informed choice. They can use the general purpose lanes for free with the likelihood of being delayed, they can car share, or pay to enjoy a hassle free and predictable journey time – or they can use the *Inland Breeze* express bus service that is funded by the tolls. While initially there were concerns that the lanes would become 'Lexus Lanes', i.e. only used by the rich, this has not been borne out in practice.

This combination of benefits for both toll and non-paying toll car drivers and public transport users is a major factor missing from many area charging schemes. There are real benefits for everyone in this type of LET.

Local motoring taxes

In most countries there are no local motoring taxes. However, in the USA, for example in Nevada, Oregon and New Mexico, it is possible for local motoring taxes to be levied by local jurisdictions for local purposes and collected in addition to State and Federal motor fuel taxes. These can generate significant revenues of which a part is often earmarked for transit objectives. This section contains examples of both excise and fuel taxes. First, the fuel tax levied in Florida will be described, followed by the motor vehicle excise tax in the State of Washington. Florida has two types of local motor fuel taxes for transport: a voted gasoline tax and local option gasoline tax. The State of Washington did have a Motor Vehicle Excise Tax (MVET) as a dedicated source of funding for local public transport until recently, while three counties there still retain a local variant. Other examples will then be examined.

In the United States, State-enabling legislation is required for local jurisdictions to levy local motor taxes. Restrictions are often imposed on the localities as to the use of the revenues, the rates that may be imposed, and the procedure for local approval of the tax. In Texas, it is possible that revenues collected would have to be distributed in the same fashion as State motor fuel tax revenue, with approximately three-quarters to transport and one quarter to education. Significant revenues can be obtained, varying according to tax rates and travel patterns. This last point may be one of the weak aspects of the scheme; revenues depend on travel patterns, which may fluctuate over time – but the overall trend is upward. Also external factors will have an impact on these, during economic recessions for example.

It is always difficult to implement a new tax in the United States, and this holds also (if not especially) for local motoring taxes. The community must accept the need for revenue and the existing motoring tax structure must not be viewed as too high to accept an additional local tax. To this extent, any potential local tax must be considered along with existing State and Federal taxes. There are some circumstances in which residents may particularly support a local motoring tax, for example in localities which have significant traffic from non-residents onto whom the tax may be passed.

Local motor fuel taxes

The advantages of motor fuel taxes are that they have a broad tax base and somewhat inelastic demand. Fuel tax therefore has the potential to raise considerable revenues, although surcharges would raise less and may be less predictable and stable. Fuel taxes are also notoriously regressive, but this is counterbalanced as they also exhibit a strong cost benefit relationship when dedicated to public transport programmes. Since most governments already have national motor fuel taxes, collecting surcharges would involve few additional administrative costs.

The US Environmental Protection Agency felt that improved fuel efficiency could cut the yield of local fuel taxes as most authorities use flat per litre/gallon rates (USEPA, 1999). It felt it might be difficult to legislate new earmarking and surcharges, and safeguard dedication to public transport. In practice this viewpoint can be challenged. As discussed in Chapter 1, improvements in fuel efficiency have always been more than counterbalanced by increased consumption and overall fuel use has increased. This, after all, is the problem driving the need for demand management transport policies.

Local motor fuel tax, State of Florida, United States[12]

Florida is one State in the USA that extensively hypothecates fuel tax revenues to fund public transport programmes, while State law allows counties within

Florida to levy fuel tax supplements too if they wish. In brief, the US Federal government levies a tax of 18.4 cents per gallon (€0.05 per litre) of which 2.86 cents (€0.007 per litre) is for mass transit. Additionally, the State levies 20.6 cents a gallon (€0.052 per litre) of which around 1 cent a gallon is dedicated to a public transport grant. This generates around $US75m (€84.3m) a year through the State Transportation Trust Fund's Transit Block.

Finally, each of the sixty-seven counties is able to levy two types of fuel tax that may be hypothecated to public transport. The first local gas tax was initially called the 'Ninth-Cent Fuel Tax' (because the State's fuel excise taxes then totalled 8 cents) and was first authorized in 1972 by the Florida State Legislature. It was renamed the Voted Gas Tax in 1983 when the State's fuel taxes increased to 9.7 cents a gallon (€0.024 per litre). The tax is limited to 1 cent per gallon (€0.003 per litre) on highway fuels, has no time limit and, until 1992, had to be approved by the electorate in a countywide referendum. The 1992 Legislature authorized smaller counties to impose the tax by a vote within their governing bodies, with the 1993 Legislature removing the referendum requirement altogether so any county can now impose the tax if its board of commissioners vote to do so. As a result, the tax was renamed the Ninth-Cent Gas Tax once more in 1996.

The Ninth-Cent Tax on diesel fuel ceased to be optional from 1994, after the 1990 Legislature decided to equalize all optional taxes on diesel fuel so that interstate truckers, who pay fuel taxes based on miles driven in the State, would be subject to standardized tax rates.

By January 2000, thirty-nine counties had implemented the Ninth-Cent Fuel Tax on petrol and gasohol. Altogether in financial year 1999–2000, $US1m (€1.1m) was spent on administration and $US1m (€1.1m) on collection, while $US58m (€65.2m) was dedicated to 'any legitimate county or municipal transport purpose'.

The second type of local gas tax in Florida is the Local Option Fuel Tax, which was introduced in 1983 as part of a restructuring of State transportation taxes. Originally called the Local Option Gas Tax and renamed in 1996, it was established as a tax of 1 cent to 4 cents on each gallon (€0.003–€0.01 per litre) of highway fuel, which could be levied by the county's governing body. While initially the tax was to be created for a maximum period of 5 years, this was increased to 10 years soon afterwards in order to make it at least minimally suitable as a security against which to issue debt. Revenue should be shared with municipalities (still the case), and the money was collected at the wholesale level along with the fuel excise taxes and the fuel sales tax.

In 1985, counties were authorized to raise the maximum rate of the tax to 6 cents per gallon (€0.015 per litre) and its duration to 30 years. Simultaneously, collection of the tax was moved to the retail level to identify better the location (and the tax rate) of where fuel was sold. Interestingly, in July 1996 the tax collection point was shifted back to the wholesaler (for gasoline and gasohol) and the terminal supplier (for diesel fuel) to make tax administration more efficient for both the State and the fuel industry.

In 1990 the State Legislature decided to equalize the Local Option Fuel Tax on diesel fuel from the start of 1991, with the minimum tax rate set at 4 cents a gallon (€0.01 per litre). This was increased by 1 cent over each of the following 2 years to 6 cents a gallon (€0.015 per litre).

At first, proceeds of the tax could only be used for transport purposes, but in 1992 it was decreed that any 'small county' could use the proceeds for other capital infrastructure needs if the transport element of its comprehensive plan has been fully satisfied. However, this exception applies only to the 6 cents of tax authorized prior to 1993. In 1993, the State Legislature offered counties the option of imposing a further 1 cent to 5 cents on each gallon (€0.003–€0.013 per litre) of motor fuel (gasoline and gasohol, but not diesel). As a result, counties may now levy a tax of up to 11 cents on each gallon (€0.028 per litre) of gasoline.

To introduce the gas tax, there are slightly different rules for the first 6 cents, and for the second 5 cents. To alter the first 6 cents of the tax on motor fuel, a majority vote of the board of county commissioners or a countywide referendum initiated by either the county commission or municipalities representing more than 50% of the county's population is required. For the second 5 cents to be charged, an extraordinary vote of the county commission or a countywide referendum initiated by the commission is needed.

As of January 2000, all sixty-seven counties had a Local Option Fuel Tax for petrol and gasohol in place, in amounts ranging from 3 cents a gallon to 11 cents a gallon (€0.008–€0.028 per litre). Income from the local option fuel tax in 1999–2000 was $US647m (€727m), of which $US587m (€660m) went to local transport projects. Of the remainder, $US13m (€14.6m) paid for administration costs, collection costs, and refunds to farmers, fishermen and transit systems, while $US47m (€53m) was paid into general revenue budgets.

Air pollution control fee, Taiwan, Republic of China

This is a notable example of a *Pollution Charge* LET which is based on the amount of pollution generated through emission charges levied on the volume and toxicity of pollutants emitted into the atmosphere. In Taiwan, money derived directly from an Air Pollution Control Fee, collected through a surcharge on fuel, is used by the Environmental Protection Agency to provide subsidies ranging from $TWD23,000 (€700) to $TWD1m (€30,800) to install pollution control devices.[13] In addition, it also pays out $TWD0.2m–$TWD0.5m (€6,200–€15,400) to bus companies for each bus they replace with a new one that meets Stage II standards (EPA of Taiwan, 1998; Chen and Fang, 1998). To avoid public misconception that such fees are a general tax, a foundation has been established to administer and manage the money collected. The foundation is also responsible for prioritizing how the money is spent. The fee was first introduced in Taiwan in 1995 under the Air Pollution Control Act Amendment of 1992, and altogether raises around $US170m (€190m) a year (including fees from non-transport sources).

Local fuel tax LETs elsewhere

In Montreal in the Province of Quebec, The Agence Metropolitaine de Transport (AMT) imposes a 1.5 cent (€0.01) per litre gasoline tax. As of the end of 2000, this contributed $C44.5m (€31.8m) to an overall budget of $C192m (Joubarne, 2001). Another Canadian example is the Vancouver Region of Canada, where in 1999 the Greater Vancouver Transport Authority Act allowed the proceeds of an 8 cents (€0.06) per litre fuel tax to be earmarked to the regional transport authority. This is double the previous fuel tax rate that was transferred from the Province for transportation purposes before the regional body was created. However, there is no overall increase in tax rates to the consumer – it is a redistribution of the existing revenues. The rate applies to all motor fuel purchased in the Greater Vancouver Region, and is projected to raise $C128m (€90m) a year (GVTA, 1999).

Finally, the Northern Virginia Transportation Committee (NVTC) and Potomac and Rappahannock Transportation Committee (PRTC) both raise money through a 2% excise tax on motor fuel sales. Most NVTC revenue pays costs of Washington Metropolitan Area Transportation Authority. PRTC funds Virginia Railway Express, highways and a bus system (Goldman *et al.*, 2001).

Local fuel and emission taxes: lessons

In general, motor fuel tax is an attractive revenue source for transport improvements. It is easily administered compared to many other taxes and provides a relatively stable revenue stream. Most important, it is paid by car drivers who are direct beneficiaries of better transport. This link ensures the tax is seen as fair. Fuel tax remains a very widely used mechanism at the national and regional levels. But, there are limits at the local level. As the tax is levied as an amount per volume sold rather than a percentage of the fuel price, generated revenues will tend to lag over time as the real value drops due to inflation. More serious is the very limited revenue base of the petrol tax. Because it only taxes one product, the rate must be set very high to generate sufficient cash to pay for major infrastructure projects.

The complications of how the local motor fuel taxes have emerged in Florida should not obscure key lessons. Here the motor fuel tax raises very large sums (although only some of this goes towards funding public transport). The tax itself may be regressive, but the combined tax-expenditure system may not be, especially for funding local public transport. It is notable

continued on page 115

---

continued from page 114

that all the variable rates at the State and county level have not produced significant 'border' effects, although this is in the general context of motor fuel taxes in the USA being low relative to Europe.

The general level of motor fuel taxation is important. Although it is not usually difficult to enact general taxes in excess of 15 cents per gallon (€0.04 per litre) for national or regional governments, at the local level, border effects are likely if highly differential rates are produced.[14] Drivers will start to buy their fuel elsewhere (Goldman et al., 2001). The Vancouver example of reallocating existing tax revenues gets around this problem.

For all their benefits, local motoring taxes in the USA and other countries do not appear to represent any attempt to charge for pollution, they are not intended to favour greener fuels (although they may do so) and certainly do not link into demand management transport planning. They are a mechanism that has not really developed into a modern transport policy LET, but have the potential to do so.

Ad valorem vehicle taxes

Motor vehicle taxes can be levied on the sale of new and used vehicles. They may include recurrent (annual or biennial) registration of existing vehicles, or registration fees may be used as a surrogate for a sales tax. Many governments charge substantial taxes for the purchase of motor vehicles, as well as ongoing registration and licensing taxes. Generally, the funds raised go either to general revenues or to pay for road-related programmes, but sometimes they are hypothecated to funding public transport (USEPA, 1999).

Motor Vehicle Excise Tax, Sound Transit, State of Washington, USA[15]

Three counties – Snohomish, Pierce and King (which includes Seattle) – of the Central Puget Sound Regional Transit Authority or 'Sound Transit' in the State of Washington have used a Motor Vehicle Excise Tax (MVET) since November 1996. This is set at a rate of 0.3% of the fair market value of motor vehicles, and in 2000 raised $US51.4m (€57.8m) out of a total income of $US290m (€325m). The Sound Transit MVET was introduced under State legislation in 1971, which enabled any municipality to adopt an MVET up to a rate of 1%, which was known as the municipality levy.

Interestingly, this same legislation introduced a State-wide MVET which operated until 1999 when it was repealed under Initiative 695. The State MVET was set at 2.354%, and cities and counties were permitted by the State to direct nearly half (1%) of this to meet local public transport needs. The remainder went to the State ferry system (0.2%) and to the State general fund (1.154%). Any entity or municipality was eligible to collect the MVET levy except for city systems that had a sales tax dedicated to transit. The MVET funds also had to be matched 1:1 using a local tax source from within a transit system's service area, or local general service fund revenues. Local tax sources included local sales taxes, or household or business taxes. Systems using MVET funding submitted budgets each year to the State Department of License which projects tax revenues, and MVET expenses were then compared with actual tax receipts, which were submitted in the April of the following year. The Department of License then adjusted current year MVET funding as needed. The MVET funds were collected by the State and disbursed quarterly with a six-month lag. This funding for local public transportation could be used for operating or capital expenses.

Rental car tax, State of Washington, United States[16]

Very closely related to the MVET in the State of Washington is the car rental tax, where a tax is levied on the value of the vehicle's rental value. Rental cars are defined as passenger cars that are rented by rental car companies to customers, without drivers, for periods not in excess of 30 consecutive days. Thus longer lease agreements are excluded.

In Washington the rental car tax was first adopted in 1992 by the State Legislature. As a result, four counties implemented a 1% levy in October of that year, while the State rate of 5.9% was in place by the beginning of 1993. A local car rental tax of up to 1.944% (2.172% in the metropolitan areas of King, Pierce and Snohomish counties i.e. the Sound Transit area) was also authorized for municipalities to spend on mass transit facilities. However, only the Regional Transit Authority (now Sound Transit) has adopted this option. This was operational from April 1997 at a rate of 0.8%.

Initially, the purpose of the tax was to replace the motor vehicle excise tax and not increase the overall burden of tax for rental car companies. Previously, the MVET applied to all rental cars located in the State, even those used only temporarily in the State for short periods of time, and no apportionment of the tax was provided to reflect the time the vehicle was actually operated in Washington. Instead, the rental tax shifted the burden directly to the customers and so better reflects actual use of the vehicles within the State.

The combined tax rate for rental car customers is quite high. For example, if the retail sales tax is included then the State, county and local tax rates for car rentals in most of King County is currently 9.7% rental car taxes plus 8.8% in sales taxes. The State and local elements of the tax are collected by just over 150 rental car companies from customers. The money is then disbursed to the transit agencies through the Office of the State Treasurer.

In 2001, the State element of the car rental tax raised $US22m (€24.7m) which was deposited in the multi-modal transportation account. In the same year the 0.8% tax imposed by Sound Transit generated $US2.4m (€2.7m) to be devoted to financing a high capacity, rapid transit system. None of the money raised from the other local car rental taxes is dedicated to transit. The State tax was previously distributed in the same manner as the motor vehicle excise tax but, with the repeal of the State motor vehicle excise tax, the receipts of the State rental car tax were transferred into the newly created multi-modal transportation account.

Other examples

Triangle Transit, which serves Durham, Orange and Wake Counties in North Carolina in the United States, is another case where both MVET and car rental taxes are dedicated to finance transit. The transit agency first adopted a $US5 (€5.6) vehicle registration fee in 1991 to fund bus and rideshare programmes. This was followed in 1998 by a 5% vehicle rental tax to fund planning and initial construction costs of a rail transit system. Also in North Carolina, the transit district in Charlotte is funded from vehicle registration fees, while the Massachusetts State MVET raised $US469.3m (€528m) in FY1998 (Goldman *et al.*, 2001).

In Canada, the Agence Metropolitaine de Transport in Montreal, generates money for public transport from an annual $C30 (€21) dedicated vehicle licence fee. In 2000 this raised $C42.5m (€30m) (Joubarne, 2001).

Ad valorem vehicle taxes: lessons

Vehicle taxes can be applied in many different ways: flat annual vehicle registration fees, annual fees based on vehicle value (or some proxy), weight, age, body type, number of wheels. There can also be taxes on vehicle rentals and leases, parking and sales. Many States have ad valorem vehicle taxes whereby vehicles are taxed as personal property as property is often taxed. These laws date from the early twentieth century when mass evasion led to States shifting collection to the vehicle registration process (Goldman et al., 2001).

In the United States, there is currently a trend to phase out ad valorem taxes, with Rhode Island, Virginia and Washington having recently rolled back or phased out their vehicle taxes (Goldman et al., 2001). Although some of the revenues from these taxes may have been earmarked to help fund public transport, this is (at best) only one of several purposes. In general, where they are permitted, local motoring taxes are simply a useful fiscal instrument and are not used as a transport policy instrument. Like local motor fuel taxes, they are a mechanism that has not really developed into a modern transport policy LET, but have the potential to do so.

Airport landing fees

Another mechanism that is closely related to tolls is a levy upon air passengers. This is only very rarely used to fund public transport. Although these charge polluters and the funds are used to support public transport, it can be questioned as to whether this is in any way environmentally beneficial. The building of faster and better transport links to an airport does not provide an alternative to air travel itself. The provision of better public transport to an airport could be viewed as no different from any other supporting infrastructure that assists the attraction of air travel. Indeed, the development of better transport links to an airport would simply make air travel more attractive and so generate even more polluting travel. Furthermore, particularly in a European context, developing fast public transport links to airports will be to the disadvantage of more environmentally friendly modes of travel, such as rail.

Overall, when considered at a transport systems level, the public transport developments funded by airport LETs may have largely negative environmental and transport policy impacts.

Airport landing fee, JFK International Airport, New York, United States

For many years, JFK International Airport, near New York City has suffered from poor access. This finally led to construction of the so-called 'Airtrain light rail link' to JFK from Manhattan beginning in 1998. Due to be completed in late 2003, the 13 km $US1.5bn (€1.7bn) project has ten stations, and is to be financed entirely by system users, the Port Authority of New York and New Jersey funds, and revenue from an existing $US3 (€3.40) surcharge on departing passengers under the Passenger Facility Charge (PFC) programme.

Using pollution charges to pay for public transport

Most forms of transport are subject to some form of taxation or charge – either on an annual or usage basis. Taxing car ownership, fuel excise and parking charges are common methods. The historical reasons for such taxation lie mainly in the need to raise revenue. They can form, and in some instances have formed, the basis of indirect fiscal instruments for environmental policy. While they are generally seen as rather blunt instruments in this latter role, the fact that most of them are already in place means that it is often politically expedient to adapt them in the best way possible. However, we have seen that there are other, more effective, alternatives such as road pricing, but acceptability seems hard to achieve for these schemes. That is why these are less common and have often failed to be implemented.

Although public transport might seem a legitimate form of spending for these revenues, road and vehicle taxation and charges are only rarely directly hypothecated to urban public transport (in particular in the United States). But, this is not grounded in transport policy principles to manage traffic or even to 'green' vehicles, but is simply one of a number of alternative local sources of taxation that is permitted under the decentralized structure of US governance.

More recently one can distinguish a trend towards more innovative plans to use money from car users to subsidize its substitute, i.e. local public transport. This stems mainly from increased policy interest in pricing the external costs (e.g. pollution, congestion) of car use (CEC, 1995, 1998). The LETs examined in this Chapter appear to reflect in some form the 'polluter pays' principle, but in practice it appears this is far from the case. Certainly in the United States, there does not appear to be any attempt to charge for pollution and the LETs do not link into demand management transport planning. In most cases it is little more than coincidental that car users are the subject of a local tax that is used to support public transport. They are really only 'spreading the burden' LETs. However, what is important is that some of these local charges, in particular circumstances, have evolved to address traffic management, environmental and other transport policy concerns. The bridge tolls in the United States are a notable example. Thus they do contain useful lessons for developing LETs that are intended from the outset as instruments of transport policy.

Parking charges, levies and fines have nearly always had some traffic demand management function, and their further development into fully fledged LETs suggests they offer great potential. The examples in this Chapter also contain lessons that apply generically to all LETs. These include:

- Careful justification, with targeted exemptions to cover equity issues – both for social reasons and for any major 'losers'. In design, it should be as simple as possible to understand.

- The public transport improvements funded by the charges clearly need to be of the order to win acceptance from business and users. Proceeds from the levies need to be seen to outweigh the charges paid.

Interestingly the seemingly idiosyncratic HOT lane example in San Diego also draws some important generic lessons. Its origins may have been far from demand management transport planning, but the outcome provides a genuine and informed choice to users. Drivers can use the general purpose lanes for free (with the likelihood of being delayed), or else pay to use the guaranteed uncongested lanes, or they can use the express bus service that the tolls fund.

This linking of introducing a LET to benefits and choice, was also highlighted in the Oslo area toll case. The package of benefits was not just for public transport, but for motorists as well. It was not an antagonistic 'motorists lose' and 'public transport gains' scheme, but one where, although changes in behaviour were expected, there were real benefits for most people. This was also true of the parking levy schemes in Australia. This links into a crucial issue for the development of LETs that are intended to be instruments of transport and environmental policy. They need to be designed to offer clear benefits for as many user groups as possible in order to achieve wide public acceptance and win the political will of politicians. Thus, even though many of the LETs considered in this Chapter have somewhat dubious transport planning credentials, they all contain important lessons for the design of successful LET measures that can be integrated into transport demand management policies.

Notes

1 Based on Lamb (1999).
2 VAT in the UK is 17.5%.
3 Based on Weir (1999) and Usher (1998).
4 Based on Reggiani (1999).
5 The figures have been converted from Dutch Guilder (NLG) to Euros at a rate of €1 to NLG2.2 (XE, 2001).
6 Based on Brown (2001).
7 The figures have been converted from Australian Dollars ($A) to Euros at the rate of €1 to $A2 (XE, 2001).
8 Based on Thoms (2001).
9 Based on Waerstad (2002).
10 The figures have been converted from Norwegian Kroner (NOK) to Euros at the rate of €1 to NOK7.5 (XE, 2001).
11 Based on Pessaro (2001), Schumacher (2001), Shreffler (2001) and San Diego State University Foundation et al. (1997).
12 Based on Florida Department of Transportation (2000).
13 The figures have been converted from Taiwanese Dollars (TWD) to Euros at a rate of €1 to TWD32.4 (XE, 2001).
14 There are exceptions. In September 2000, the British Government was forced to abandon its 'fuel tax escalator' due to significant public opposition (Lyons and Chatterjee, 2002).
15 Based on Washington State Department of Transport (2000).
16 Based on Washington State Department of Revenue (2002).

Chapter 5

Spreading the burden: LETs to meet social objectives

Maximizing revenue with minimum controversy

The final group of local earmarked taxes (LETs) – those in the 'spreading the burden' category – stand somewhat in contrast to the *Beneficiary* and *Polluter Pays* categories. There is no connection between the source of the income and the purpose to subsidize public transport services. They neither seek to charge those who benefit from improved public transport services nor provide a financial discouragement to transport polluters. Instead, as discussed in Chapter 2, these measures are essentially designed to provide as broad a tax base as possible. Any link to other principles of public finance or transport and environmental policies appear coincidental.

The rationale for public transport subsidized by these LETs tends to be that of social inclusion and equity. Yet so separate is the gathering of income from its use, that the issue of the distributional effects of these taxes and charges to pay for this socially-led policy does not feature in any significant way. In general, the use of such 'spreading the burden' measures has occurred where 'traditional' finance sources have failed to keep pace with the funding needs of public transport systems. But, rather than designing a LET to influence travel behaviour or to target a group of beneficiaries, this group of LETs has emerged with the specific intention of avoiding any kind of 'philosophical baggage'. The whole idea is to have a good, uncontroversial fundraiser. Because of a number of special circumstances, the vast majority of examples of this type of LET are to be found in the United States. Crucially, outside of New York City public transport tends to be used only by a relatively small minority of people and most citizens are therefore unhappy paying, through general taxation, for something they never use. On the other hand, many (even American) people recognize the wider social, economic and environmental benefits of providing public transport – not least because it shifts the other guy out of his car. This perception, coupled with the existence in the majority of States of mechanisms for localities to raise additional monies for specific services has meant that quite a few local authorities have been able to develop at least a basic level of public transport service. In Europe, until recently at least, the higher level of public transport use and the facts that driving a car is more expensive and roads are more congested than in the United States has meant that most people have been prepared to fund public transport through general taxation. Moreover, there is no real history of local authorities raising large sums of money through locally dedicated taxes.

Taxes on consumption can take several forms. This section considers the following types of locally earmarked consumption taxes:

◆   general sales taxes;
◆   sales taxes on specific items (e.g. beer, cigarettes, hotel rooms);
◆   amusement taxes (e.g. gambling, lotteries);
◆   utility levies.

It also briefly examines the use of cross utility financing mechanisms. These are not strictly LETs, but are closely related and so are included for completeness.

There are many examples of general sales tax LETs in the United States. The specific design varies somewhat according to the transit agency involved, with different modes, locations, and circumstances. This Chapter discusses only a small number of examples of general sales taxes, namely a small system in Reno, Nevada and a large multi-modal system in Atlanta, Georgia. In addition, three examples of a specific sales tax are considered. These are: the beer tax implemented in Birmingham, Alabama, a cigarette tax used in Oregon, and taxes on hotel rooms in various locations in the US.

Amusement taxes are also used to fund public transport. For example, in Atlantic City, New Jersey taxes from casinos are earmarked to public transport while proceeds from lottery taxes are dedicated to transit in Pennsylvania and Arizona.

Finally, this Chapter will cover two examples of earmarked utility levies (i.e. taxes on the use of services such as electricity, gas, water etc). The first is in the US City of Pullman, Washington, and secondly a method of cross-utility financing (more of an internal cross subsidy than a tax or levy as such) in Wuppertal, Germany.

Local general sales taxes

Local general sales taxes are the most widely used earmarked charge for funding public transport in the United States. This is because revenues from these taxes are broad based, sizeable and relatively stable. Yield will drop in a recession, but so would the yield from normal general sources of taxation as well. They are also relatively uncontroversial because they are levied across such a wide range of products at such a low rate that people do not notice they are paying as much tax as they are.

Local sales taxes are typically an additional rate on existing State general sales and use taxes, but in some cases they have also been introduced where there is no State sales tax. Depending on State constitutions, statutes and home rule traditions, most local governments must seek State approval to levy local sales taxes, as well as local voter approval. Once authorized by the State, local taxes are usually limited to a specified time period, or a maximum collection total, and a specific use. The dedicated revenue stream may be used directly to fund a public transport service, or it may service a loan, such as a local general obligation or revenue bonds, which might have been raised to pay for a capital project.

Although the revenue may be used to fulfil a social view of the role of public transport, sales taxes are inherently highly regressive. There are also practical difficulties. States and localities may have statutory limitations on general sales tax increases, the local approval of tax increases may be time-consuming and is not assured, and earmarking can be difficult to sustain (USEPA, 1999). Financial dependence on such a LET can also produce some negative transport effects. There have also been charges that local authorities

Figure 5.1 As in many US cities, Chicago uses general sales taxes to support its public transport services

have encouraged or even provided incentives to retail development to increase the tax take. The emergence of so called fiscal zoning in California is one example of this phenomena that has attracted widespread attention, with charges that it distorts both the market and the process of land-use planning (Teitz, 1999). Thus the very measure intended to finance public transport potentially ends up encouraging car serviced sprawl, difficult and costly to serve by public transport, and undermining the purpose for which the LET exists!

Local general sales tax, Washoe County, Nevada, United States[1]

One example of how local general sales taxes have been earmarked to pay for public transport is in Washoe County, Nevada, where a transit system began operating in 1978. Initially this relied on two revenue sources: fares and subsidies from the general budgets of Washoe County and the cities of Reno and Sparks, but in 1982, the cities and the county began to have monetary problems. As a result, a 0.25% sales tax was adopted to provide the necessary revenue source to pay for general transit and specialist services for the elderly or those with disabilities.

In Nevada, the State legislature must authorize all local tax proposals, before they are voted on in a local referendum. In order to implement the 0.25% sales tax to pay for transit and road improvements, several key actions were undertaken to generate community support. For example, a proactive community outreach programme was carried out where transit representatives spoke to service clubs, businesses, and other members of the community about the benefits of the transit system. In the event, the sales tax in Washoe County secured the approval of 70% of the electorate.

Proceeds from the sales tax go to an account for the Regional Transportation Commission of Washoe County (RTC) and RTC gives the county treasurer permission to invest it with other unused county funds. When the RTC wants to use the sales tax revenue, it draws the money from this account, deposits it into its own transit expenditure account. In 2000, the operational budget of RTC was around $US18.1m (€20.3) of which the sales tax provided just over half, at $US9.2m (€10.3). Fares covered $US5.5m (€6.2m), while Federal assistance amounted to $US2.8m (€3.1m) (National Transit Database, 2001).

Local general sales tax, Atlanta, Georgia, United States[2]

The Atlanta example provides a rather larger application of a dedicated sales tax.

In 1965, the State of Georgia created a transit authority to serve metropolitan Atlanta, the Metropolitan Atlanta Rapid Transit Authority (MARTA). While the first referendum to create a five-county rapid rail system failed in 1968, in 1971 the City of Atlanta, Fulton County and DeKalb County approved a referendum for a 1% sales tax to support a rapid rail and feeder bus system. Cobb County and Gwinnett County voters rejected the MARTA system. In 1972, MARTA acquired the Atlanta Transit System, which was a private company, and began operating bus services.

The sales tax is sent to Georgia's State Revenue Commissioner each month. The Commissioner takes from this MARTA's monthly debt service payments and turns over the remaining money to MARTA. The State of Georgia charges MARTA a handling fee of 0.5% of its total sales tax receipts to cover the collection costs. Merchants also keep a portion of the tax revenues earned by MARTA, this decreases the total potential receipts for the transit organization. Half of this revenue must be spent on the operation of the transit system, while the other half is to be spent on the construction of infrastructure. Once again the sales tax provides the majority of MARTA's funding and has allowed the development of an extensive heavy rail system. The sales tax revenue has historically provided over half the operating funds, while passenger fares cover just under 40% of the operating revenues. For example, in 2000 MARTA's operating cost amounted to $US339m (€380m). Of this, passenger revenues covered $US95m (€106m) while local taxes (of which the sales tax is predominant) provided $US212m (€237m) (National Transit Database, 2001).

From 2002, the sales tax rate was due to fall from 1% to 0.5%. This will cause some problems in covering the operating expenses, although by that time, MARTA expects to have completed the construction (as planned in the 1970s) of its system and paid off all the associated bonds. In addition the proportion of the tax revenue spent on operating subsidy will be increased to 60%.

Other general sales tax experience

General sales taxes are used in other US cities, such as Austin in Texas, where a 1% transit sales tax is levied raising approximately $US7m (€7.8m) per year. Dedicated sales taxes are also used to finance public transport in India. The State Government of Karnataka for example, introduced a sales tax to finance a light rail line in Bangalore. In Mumbai meanwhile, a 5% surcharge is levied on the sale of selected goods and services to finance the construction of the city's seventh rail corridor (discussed in Chapter 3). The General Metro Tax is expected to yield Rs1bn (€23.5m) a year, and will be discontinued after the capital investment of the project has been recovered, unless the government wishes to extend the line to other locations (Dalvi and Patankar, 1999). In Teito, Tokyo, there is an instance where a department store constructed on rail land paid a subsidiary of the undertaking a contribution based on the level of sales (Ridley and Fawkner, 1987), although this is not strictly a sales tax.

While general sales taxes are apparently politically acceptable in many areas, there are exceptions. For example, a proposal for a local surcharge on Value Added Tax (VAT), dedicated to funding a public transport system in Madrid, Spain, was soundly rejected politically, and led to its proposer's downfall (Farrell, 1999*b*).

General sales taxes: lessons

Overall a general sales tax LET has the advantage of tapping into a broad tax base. It has a high income potential and because it is not linked to an individual product or sector, revenues should be reasonably reliable (but income will be affected by the general economic cycle). Like many consumption taxes, a general sales tax itself is regressive (although the combined tax/expenditure system may not be) and there can be difficulties maintaining the earmarking to public transport. Where a general sales tax already exists, a supplementary LET can use existing administrative arrangements. Where there is no existing tax, gaining approval and setting up a new administrative system is a disadvantage.

Selective sales taxes

Selective sales taxes on the sale of particular commodities are levied either as a percentage of the sale or price of the item, or as a flat charge per item. Selective sales taxes can be more easily dedicated to a particular public transport programme compared to general sales taxes, since there often is a more direct relationship between the particular type of product in the tax base and the use of the funds for public transport purposes. The downside is that the tax base for selective sales taxes is much narrower than for general taxes. Therefore, a higher rate must be charged to generate the same amount of revenue, which may cause inequities. General sales taxes typically are highly regressive, since it is difficult to use graduated rate structures depending on the economic circumstances of the purchaser, although sales taxes on luxury items are less so. While State use of selective sales taxes in the United States is widespread, revenue yields remain modest (USEPA, 1999).

Interestingly, the most significant local selective sales tax for financing public transport, is the petrol tax. This is actually reported on in the previous Chapter under the 'Polluter Pays' category (although in many cases it may be little more than coincidental that it was petrol that was the product taxed!). Examples covered in the following section are beer taxes, cigarette taxes, and hotel taxes.

Beer tax, Birmingham, Alabama, United States[3]

One example of a sales tax on a specific product is the beer tax in the City of Birmingham (Jefferson County, Alabama). Although all fifty US States, many localities, and the US Federal government (as well as many governments world-wide), levy taxes on over-the-counter purchases of all alcoholic beverages, Jefferson County is currently the only local authority to earmark alcohol tax receipts to fund public transport.

This LET was established in April 1982, after a beer tax was introduced throughout Alabama. However, each county divides its portion of revenues from this tax differently, and only in Jefferson County are beer tax receipts dedicated to public transport. The beer tax, of 1.625 cents for each 4 fluid ounces of beer (€0.0014 per litre), is collected by the assessing authority of the county or municipality. In Jefferson County, the money is deposited in three different funds, (after 2% is removed for administration costs by the County). Altogether, a proportion of the third fund (Fund C), which represents one-third of the tax received, is distributed to the Birmingham-Jefferson County Transit Authority. This currently amounts to 50% of Fund C, or $US2m (€2.2m) annually, whichever is greater. Revenues from the beer tax have amounted to 17.8% of the Transit Authority budget each year, and have been used mainly for capital expenditures.

An advantage of a tax surcharge on beer to fund public transport is that it is administratively simple to collect and track because administrative records of alcohol sales already exist. Consumption is widespread, and so revenues could be significant with a relatively small additional tax. The demand for alcohol is relatively inelastic, meaning a tax increase may not cause a decrease in sales. Indeed, in times of economic recession, beer consumption tends to be little affected! A specific sales tax on a product with reliable demand makes for a reliable income stream, but, as for most sales taxes, the beer tax is highly regressive (USEPA, 1999).

Cigarette tax, Oregon, United States

As with beer taxes, local cigarette taxes are levied across the United States, although only two examples were found where receipts were dedicated to pay for public transport, and one (in Massachusetts) no longer operates. Legislation for the Oregon cigarette tax was passed by the State Legislature in 1999. This states that 3.45% of a State cigarette tax 'shall be appropriated to the Department of Transportation for the purpose of financing and improving transportation services for elderly and disabled individuals' (Oregon State Legislature, 1999a). In 2001, the Oregon Department of Transportation (2001 distributed $US8m (€9m) across the State (Oregon Department of Transportation, 2001). In 2001, the State Cigarette Tax contributed $US1.5m (€1.7m) of the Tri-Met (Portland Region) operational budget (Tri-Met, 2001).

Hotel room taxes, United States

Hotel taxes are levied on rooms, or guest occupancy, charged either per night or as a percentage of the room rate. The major advantages of occupancy taxes are that they spread the costs of maintaining services to those who benefit from them and who would not normally contribute, i.e. to non-local 'residents'. They are therefore equitable, and relatively popular with the local electorate (who do not have to pay the tax), but are obviously less popular with the tourist trade. To some extent hotel room taxes could be viewed as a 'beneficiary pays' LET (if tourists are users of public transport), but it does not appear that this principle really underlies the use of this LET in practice.

A disadvantage of a hotel room LET is that, because the demand for hotel space is relatively elastic, a price increase may reduce occupancy rates, and therefore tax revenues, particularly if a city or county unilaterally imposes an occupancy tax higher than in surrounding areas. Setting up a hotel room tax from scratch may also involve high administrative costs if no occupancy tax currently exists. In summary, revenue yields may be low, unpredictable, and lack stability (USEPA, 1999).

In Louisiana, New Orleans Regional Transit Authority recently won approval to collect a 1% hotel tax to help pay for local streetcar capital improvements. It is expected to raise $US4.2m (€4.7m) a year and will go towards meeting the RTA's 17% contribution to the $US157m (€176m) project (Eggler, 2001).

Figure 5.2 Traditional Streetcar in New Orleans

In Nevada, a 1% lodging tax is used to part-fund a railway grade separation project in downtown Reno, while in Oregon a hotel and motel tax collected in four jurisdictions is earmarked to public transportation projects (Goldman *et al.*, 2001).

Selective sales tax: lessons

Overall a selective sales tax LET depends on a narrow tax base and, as such, raises limited funds. There is a danger with income being very dependent on the sale of the good or service concerned, and this is unlikely to vary with the need to support public transport. Sales of goods such as beer and cigarettes are, however, notably stable. However, selective taxes are regressive (although the combined tax/expenditure system may not be). As with the general sales tax, where there is no existing tax, gaining approval and setting up a new administrative system is a disadvantage. It may not be worth it for a relatively small and erratic revenue.

Amusement taxes

Amusement taxes can be levied on a wide range of activities including attendance at sporting or entertainment events, gambling and lotteries. The benefits to a tax authority are that they spread the costs of providing services to visitors from out of the locality, who might benefit from using public transport when attending an event for example. A further plus, is that ticket sales are relatively easy to track, although government collection systems must be established. However, revenue yield may not be high, while demand for tickets to sporting and other entertainment venues can be relatively sensitive to price increases. Therefore taxes could reduce the number of tickets bought and thereby lower revenues.

When searching for examples of local amusement taxes, none were found where taxes on ticket sales were earmarked to pay for public transport. But, an example of a gambling LET was found in Atlantic City, while lotteries were found to pay for public transport in Pennsylvania, Arizona and Oregon. These are reported below.

Gambling tax, Atlantic City, New Jersey, United States

In Atlantic City, New Jersey, each casino is taxed 8% of its gross revenue (the amount the casinos keep after all bets are paid). The tax is deposited in an account known as the 'Casino Revenue Fund'. The money is then distributed to authorized programmes throughout the State (Casino Revenue Fund Advisory Commission, 2002). The tax was introduced in 1976 following a voter referendum to permit gambling in Atlantic City. This specified that revenue from the tax on casinos must be used to fund health, transit, and social programmes for the aged and disabled.

Casino revenue funds are allocated to each county for local para-transit service including door-to-door and fixed route services and also subsidizes bus and rail fares for the elderly. In fiscal year 2001 a total of $US24.5m (€27.4m) (or 6% of the revenue raised from the tax) was spent on transportation programmes.

Lottery income, Pennsylvania, United States

Lotteries sell tickets for a chance to win a sum of money or other valuable prizes. Where operated for the benefit of State or local government, they generally retain a portion of the revenue from ticket sales. In the United States this proportion ranges from 10% to 50% depending on the game (USEPA, 1999). In general the revenues produced from lotteries are used for a variety of 'good causes' which have to bid to a lottery fund, but in some circumstances part of the income from a lottery is earmarked to specific causes. In some cases this includes public transport.

The Pennsylvania Lottery was established by the State Legislature in 1971, but it was not until 1972 that its first game went on sale. This delay in implementing the tax by the Bureau of State Lotteries was due to the complexity of establishing procedures for the games, the rewards, and the distribution network of retailers who sell lottery tickets. The State Department of Ageing, the Department of Transportation, and the Department of Revenue were also involved in setting up the levy, the primary purpose of which was, and remains, to generate funds for benefit programmes for the Commonwealth's older residents (Pennsylvania State Lottery, 2002).

The transport element is administered by the Pennsylvania Department of Transportation. This allows persons of 65 years of age or older with a proper ID card to use the Shared Ride programme – which offers door to door specialized transportation services (vans and mini buses) – at a reduced fare throughout the State.

In fiscal year 2000–2001, the Lottery achieved sales of approximately $US1.8bn (€2.0bn), and total programme contributions were over $US626m (€701m) of which the transportation share was $US114m (€128m).

Lottery tax, Arizona, United States

Another case where State lottery receipts are earmarked to fund public transport, is in Arizona. The Arizona lottery was established as a result of a citizen's initiative passed on November 4, 1980. During the financial year 2001, the lottery raised $US79m (€89m) for State-wide funding programmes. Of this, the Local Transportation Assistance Fund (LTAF) received $US23m (€26m), while the Mass Transit Fund received $US3.7m (€4.1m) (Arizona State Lottery, 2002). The funds are distributed to cities and towns on the basis of population (Arizona State Lottery, 2002). The funds can be used for public transportation and other transportation purposes depending on the jurisdiction's population. This fund is not administered by the Arizona Department of Transportation.

The 1998 Arizona Legislature passed State Law HB 2565 to provide additional State-wide transit and transportation funding to incorporated cities and towns as well as the counties. The LTAF II funding is in the form of multi-State lottery game and instant bingo game monies along with a portion of the State Highway Fund's Vehicle License Tax revenue. The Department of Transportation administers the LTAF II and the State Treasurer's Office distributes the funds to the Regional Public Transportation Authority (RPTA), Metropolitan Planning Organizations (MPOs), and cities and counties not represented by a RPTA or MPO.

There was a further scheme in Portland, where in 1994 the Oregon Department of Transportation issued $US97m (€109m) worth of bonds to finance a light rail extension, which are backed by the proceeds of the State lottery (Vacarre, 1996).

Consumption taxes: lessons

Overall, all types of sales taxes can often piggyback on an existing revenue collection system. Choosing what to tax can be important, and 'vices' such as cigarettes, beer, and gambling can provide reasonably stable income flows. There is an ethical dilemma in that high tax on some of these products is intended to cut (or even eliminate) their use, yet the tax pays for a public good – public transport. Although not a 'polluter pays' LET, some selective sales taxes exhibit similar characteristics and dilemmas.

Some products can yield a higher income and bear a higher tax rate than others. The better ones can generate large, relatively stable revenues.

Lotteries are also seen as a very controversial source of revenue, with critics of the lottery pointing to the sins of gambling, the opportunities for corruption, and the high rate of participation by the poor – producing a highly regressive system.

Utility levies

Utility taxes are surcharges on regular customer utility bills, such as electricity, heating oil or gas, and even telephone charges. The advantages of taxing utilities are that they can be readily estimated and tracked on a national, State and local basis, and are easy to collect through regular billings, which the utility company would then pay to the relevant governmental unit. In addition, even low-level increases to annual residential costs for total utility use would yield a significant and relatively stable revenue source.

However, the impact of utility levies on some residential customers could be high within an already highly regressive cost structure (USEPA, 1999). The influence of external factors such as economic conditions and social trends also strongly influence tax receipts. Firstly, the utility rates themselves determine the revenue received, while secondly, because the use of utilities is fairly constant, if the rates are not raised to keep pace with inflation, tax revenue will stagnate. Another factor is resource conservation: utilities such as gas, electricity and water are used less under more aggressive conservation programmes and, thus, generate less revenue.

Utility levy, Pullman, Washington, United States[4]

One well-known example of a dedicated utility levy was established in Pullman, Washington State.

During the 1970s, the oil shortage, combined with the dearth of parking in the Washington State University area, persuaded the City of Pullman to introduce a transit system. While the State of Washington allowed a local 0.2%–0.3% sales tax to be levied as dedicated revenue sources for transit, this was not felt to be acceptable, because the city is only 11 km from Moscow, Idaho, which has lower State sales taxes, property taxes, and wages. As a result, the State legislature allowed Pullman to ballot residents on a utility levy, which was approved by voters in November 1978. The 2% levy on telephone, water and sewer (owned by the city), electric, gas, and garbage utilities was introduced in January 1979, and the public transport system began operating in March 1979.

The levy is collected by the utility companies, and transferred to the City of Pullman, which then transfers the monies to the transit department of the city. The levy pays 40% of operating costs of the city's 14-vehicle, fixed-route and para-transit service. Initially, the Transit Department borrowed $US150,000 (€168,000) from the City Street Department to buy equipment and pay wages, and the utility tax and other revenues were then used to pay back this loan. The utility levy is matched 1:1 with money from Washington State sources.

In 1984, voter pressure led to the 2% rate being lowered to 1.5%. But, the City Council did not want service cuts, and so raised it back to 2% in 1989. Unfortunately, it took until 1992 to recover financially from the tax cut and return to 1984 service levels. For the 1999 budget, fares accounted for $US388,546 (€435,000 – 26%); other local taxes accounted for $US61,128 (€68,000 – 4%); State motor vehicle excise taxes accounted for $US505,379 (€566,000 – 33%); the 2% local utility tax accounted for $US505,379 (€566,000 – 33%); and a federal grant accounted for $US49,356 (55,000 – 3%). In 2001, the utility tax was estimated to have raised around $US556,000 (€623,000), so the utility levy covers a large and increasing amount of costs (City of Pullman, 2000; LTC, 2001).

A disadvantage of the scheme is that utility rates determine revenue, so if utility rates are not raised with inflation, then transit revenue stagnates.

There is also a risk that utility prices may fall (as has happened in deregulated markets) and that energy conservation programmes reduce revenue. On the plus side, the mechanism is simple to understand, and the money is easy and cheap to collect.

Other utility levies

A levy on electric power sales is also used to fund public transport in several other US cities, such as Springfield, Missouri and New Orleans, Louisiana (Cervero, 1983). The public transport shortfall of the Briham Electric Supply and Transport Undertaking (BEST) in Mumbai (formerly Bombay) in India was subsidized by the Electricity Supply Division to the tune of Rs1463.7m (€34.4m) in FY1999/2000 (Pattison, 2001).

In Canada, residential households in the Greater Vancouver Region are required to pay a hydro levy of $C1.90 (€1.35) per month to pay for transit provision. The GVTA Act does not permit any increase in this rate, and it does not apply to non-residential electricity accounts. Around $C11.8m (€8.4m) was raised in 1999/2000 through this mechanism (GVTA, 1999).

Another example is that transport and transmission companies (including trucking and local telephone companies) operating in the Metropolitan Commuter Transportation District of New York must also pay a surcharge on their State franchise tax to support transit operations. This generated $US601m (€673m) in FY 1997–1998 (Goldman et al., 2001).

Utility levies: lessons

Overall, utility levies have the advantage that they represent quite a wide tax base and a small levy can yield significant income. They are also a reliable income source. However, like all consumption taxes, yield depends on utility prices and the amount sold. Modern trends towards deregulation of utilities seem set to cut prices, cutting tax yield, which would be further cut by energy and water conservation measures.

Cross utility financing

In Germany, public transport systems are still often municipal departments, and as such are often subsidized by revenue from other municipal departments, such as water, gas and electricity, that generate a revenue surplus. This effectively allows the municipality to offset any profits against the losses of the transport undertaking meaning that these profits are not subject to corporation tax. But, the previously profit-yielding companies are no longer as lucrative, and in the longer term the deregulation and liberalization of utilities in the EU will render such models impossible (Felz, 1992; Copenhagen Transport et al., 1995). In spite of this, Farrell (1999a) found that around 100 of the

Figure 5.3 The famous Wuppertal Monorail, which benefits from cross utility financing

174 public transport companies that are members of the transport operators' association Verband Öffentlicher Verkehr (VOV) still supply utilities, and as late as the early 1990s around 18% of transport operating costs were covered by profits from these other activities.

Cross utility financing, Wuppertal, Germany

One example of this arrangement occurs in the Municipality of Wuppertal. Here, the Wuppertaler Stadtwerke (WSW) is a public utility company, responsible for public transport, gas, water and energy supply. In addition to the tax benefits mentioned earlier, cross-utility financing is practised because by internalizing the public transport subsidy it is able to retain more independence and continuity in the services it offers with less regard to temporary public financing problems.

Other cross utility financing schemes

Other German cities including Dortmund, Mannheim, and Münchengladbach, are also partly financed by cross subsidies from municipally-owned gas, electricity and water companies (Pattison, 2001), while similar arrangements are in place in some Austrian and Italian cities (e.g. Milan), as well as in Luxembourg (Bushell, 1994; Pucher, 1988; Farrell, 1999b).

Conclusions

'Spreading the Burden' LETS have no intended transport or environmental policy link, other than the income they generate being used to fund public transport. Their use is simply to provide a regular and politically acceptable source of income. Sales taxes tend to be more acceptable than most other forms of local taxation. Although sales taxes tend to be regressive and the services they finance do not generally benefit those who pay the taxes, they do provide a stable source of revenue and respond quickly to changes in overall income levels. However, it is important to take into account the total level of sales taxation as well as the additional LET. The total amount of tax can affect political acceptability and can lead to adverse local border effects.

Local sales tax can provide a dedicated funding source for a public transport agency, and through its implementation, an agency can collect a substantial amount of revenue for system operating and capital costs. Revenues are stable and can be counted on from year to year, unlike an annually appropriated source. This system is transferable, and sales taxes are the most common locally dedicated revenue source for transit systems in the United States. Sales taxes require a strong local retail base to be an effective funding source. It would be problematic to introduce them to a fragile or declining retail base. The unique US system of voters' approval can be difficult. Even in situations where the approval process does not require referenda, careful attention needs to be paid to political acceptance in design of a sales tax.

But there are problems with most 'spreading the burden' LETs. External factors can impact on receipts. For example, a recession as in the early 1990s caused a significant loss of local sales tax revenues to Washoe County's RTC, meaning that it also lost the matched federal assistance. Between 1989 and 1992, the Metropolitan Atlanta Rapid Transit Authority's sales tax receipts grew more slowly than expected due to the recession. A feature of general sales taxes, indeed of all consumption tax LETs, is that the yield will vary not with the need for public transport finance, but with patterns of consumption for unrelated products.

Specific sales taxes earmarked to public transport are less common than general sales taxes, with the possible exception of petrol (discussed in Chapter 4). Taxes on beer and cigarettes are highly regressive but very stable sources of revenue as they are not very price sensitive. Hotel room taxes are more price sensitive but are arguably more equitable. This is because they are levied on visitors to the area, with a likely bias towards higher income groups.

Examples of amusement taxes hypothecated to public transport were only found in the United States. Gambling taxes and lotteries are also seen as being very controversial sources of revenue. It would probably be unacceptable for a lottery to be specially set up to fund public transport alone, although restructuring existing lotteries to earmark some income to public transport is viable.

Utility levies have not been particularly widespread, but where they have been used they have delivered significant, stable and easy to collect revenue. But, they are often regressive and incomes risk being substantially reduced if utilities are used less due to resource conservation programmes. For many years, cross-utility financing in Germany and elsewhere has worked well, delivering high quality public transport systems. However, with the liberalization of the utilities sector across the European Union it is expected that this source of revenues will decrease.

Overall, as was noted in Chapter 2, the rationale for public transport subsidized by these LETS tends to be that of social inclusion and equity. Revenue is being raised to provide public transport for poorer and disadvantaged groups. But in some cases, other factors have started to play a subsidiary role. For example there has also been a shift to considering environmental concerns in the justification for subsidy of some public transport systems supported by 'burden sharing' LETs. This raises the possibility that, although this group of LETs started off with little more than pragmatic income sources, they could evolve into a different form. In terms of transport and environmental policy and economic/fiscal theory, the 'spreading the burden' LETs look pretty bad, but in practice they work reasonably well. It may be that pragmatism may not be a bad starting point. This group of LETs could well be developed from their pragmatic base into a more rounded and integrated set of policy instruments.

Notes

1   Based on TCRP (1998).
2   Based on TCRP (1998).
3   Based on Rice Center (1986) and Birmingham Jefferson County Transit Authority (2002).
4   Based on TCRP (1998).

Chapter 6

Evaluating LETs

Introduction

In most countries, support for public transport has traditionally been financed from general taxation. This means that there is no direct link between the source of revenue and what it is used to finance. There is no earmarking (hypothecation) of revenue for any particular purpose. The result is that, in the competition for funds, public transport often falls behind spending for other public services such as education and health. This is problematic because transport investments often require large sums of money over long periods of time, but politically it is easier to 'make do and mend' on public transport than on health and education. This situation has led to a search for new sources of funding, which have included the private sector, via privatization or contracting agreements, and earmarked charges or taxes to provide an assured income to support public transport operations and investment.

The previous Chapters identified a wide variety of schemes where public transport has been financed by a local tax or charge. The cases reported in this book have been grouped into three categories: beneficiary pays, polluter pays, and spreading the burden. In fact, this is a rather broad and general categorization. When we look in more detail at the various schemes we can make more specific groups of cases (although some overlaps do occur). These are shown in Table 6.1.

Table 6.1 Typology of local taxes or charges

Employer/employee taxes Property (related) taxes Developer levies	Beneficiary Pays
Parking charges and fines Charges for the use of road space Local motoring taxes	Polluter Pays
Consumption taxes Utility levies	Spreading the Burden
Airport landing charges Student fees	Miscellaneous

The airport landing charges and the student fees could not be directly clustered under one of the new headings, so they were given an additional category to include them in the analysis. Starting from the above typology this Chapter will provide an evaluation of the various LET measures by drawing out common themes, lessons and experiences. The criteria used relate to key practical and policy-making aspects. In particular there is emphasis on the four key requirements of revenue raising potential, practicality, transferability, and acceptability. It should be stressed that the aim of this assessment is not to identify any overall best category. This is because success depends heavily on local circumstances (e.g. existing tax structure, institutional and legal frameworks, public acceptability). The criteria are intended more as a structure to help match an appropriate LET to a particular situation.

Given that the case studies are largely qualitative, standard statistical tools are not appropriate. Use has therefore been made of a type of artificial intelligence method, called 'rough set theory'. This enables a sectional analysis to be carried out on the case study material. It needs to be mentioned that not all cases reported in the previous Chapters have been included in this rough set analysis. On the other hand, we include cases here that have not been mentioned elsewhere in the book. The original study report (Van den Branden et al., 2000) contained information useful to our purposes here. Some failed schemes (e.g. Hong Kong, Cambridge) for instance, have been included in this analysis because these are relevant for drawing some lessons and experiences. But, only those schemes that provided us with satisfactory information enabling us to carry out this type of analysis have been included. This resulted in thirty-six suitable schemes being used.

Evaluation of LET schemes

Before evaluating the LET schemes it is useful to summarize them and to identify their principal features. Table 6.2 provides an overview that acts as a guide in the following discussion of schemes.

The following uses the assessment framework to structure and analyse the major lessons and conclusions on the various LET schemes.[1] Firstly, what can be said of the practical issues of their revenue raising potential, practicality, transferability, and acceptability?

Typically the fund raising potential appears to be high for most categories in the typology, and LETs can often form a substantial part of the operating budget or contribute in a significant way to the construction of new infrastructure (see Table 6.3). The rest of the funds normally come from fares and other (conventionally-funded) public subsidies.

On the issue of practicality, most of the examples rely on existing legislative structures, which keep costs and complexity relatively low. Only some of the

Table 6.2 LET measures and their principal features

Type of scheme	Category	Principal features
Employee/employer taxes	Beneficiary pays	usually a local charge per employee; sometimes banded with highest payments in areas of best public transport; sometimes relief for employers who provide public transport support to staff.
Property taxes	Beneficiary pays	tax upon property in areas of public transport; "user pays" concept: intended to capture some of the rise in property values generated by public transport; usually earmarked business tax; often used to pay loans/bonds.
Developer levies	Beneficiary pays	can be applied in a variety of ways, including by private developers; often linked to planning permission.
Parking charges and fines	Polluter pays	applied by both private and public authorities; makes use of existing powers.
Road space charges	Polluter pays	includes tolls, congestion and road user charges; may require new powers; can raise large sums.
Local motor taxes	Polluter pays	includes local levy on fuel and excise taxes.
Consumption taxes	Spreading the burden	local taxes on a variety of consumption goods and services; may be a general goods/services tax or on a particular good (e.g. beer or gambling); used extensively in the US.
Cross utility financing	Spreading the burden	where multi-utility companies provide a subsidy to public transport from their other operations.
Other	Miscellaneous	rest category including airport landing charges and student fees to pay for public transport.

more recent LETs (e.g., road and congestion pricing) have required new powers. This tendency to rely on existing legislative structures has implications for flexibility and transferability. It may result in the 'easy' mechanism being implemented rather than what is appropriate. Furthermore, with many LETs being implemented under country or local-specific powers, transferability can be very limited. Even if local circumstances and institutional aspects may be similar, it appears that it is not that simple to copy successful examples. It should not be forgotten that certain categories (e.g. local motoring taxes and consumption taxes) are very much a product of conditions and taxation systems prevailing in the United States. Existing institutional structures (organizing referenda) and generally low tax levels make it possible to implement more

Table 6.3 Fund raising potential of LETs

Category, case	Share in operating budget (annually) or investment
Employer tax, Versement (France)	Funded on average 33% of the budget of transport companies (e.g. 20% of RATP budget in Paris)
Employer tax, Portland (US)	Funded 56% of the operating budget of the local transport authority in 1985
Development levies, San Francisco (US)	Funded in 1996 about 2% of the operating budget of the municipal railway (Muni)
Parking charges, Heathrow (UK)	Funded 0.3% of the total expenditures of the airport (including large infrastructure projects)
Parking charges, Amsterdam (Netherlands)	In total parking revenues will fund 1.5% of the total infrastructure costs of the IJtram
Charges for the use of roadspace, Golden Gate Bridge San Francisco (US)	Funded 49% of the operating budget of the bus and ferry organization in 1997
Consumption taxes, Washoe County, (US)	Funded 50% of the operating budget of the public transport company in 2000
Consumption taxes, Fort Worth (US)	Funded 71% of the operating budget of the public transport company in 1996
Utility levy, Pullman (US)	Funded 40% of the operating costs of the local transport company

easily new taxes to fund public transport. However, one should not forget that public transport funding is very much dependent on the political environment and market developments (such as demographics (and thus demand) patterns). Travel in the United States is growing in market segments that are difficult or costly to serve by traditional public transportation and hence very much automobile oriented leading to low public support for new taxes to fund subsidy needing public transport. There may be more support in Europe for this reason. But implementing these mechanisms in Europe may also be more difficult, due to the lack of such processes and existing structures, as well as the already relatively high taxes on fuel. In the end, it may also be possible to combine funding methods, especially when both measures can work together leading to additional benefits. Parking charges, for instance, together with road tolls may be part of a certain programme to decrease road use, and offer road users an improved alternative (public transport).

The issue of public acceptability is perhaps the most difficult one to address. Often, acceptability is low when a new charge or tax is introduced, but improves when the objective (i.e. to fund public transport) is explained and the new or improved services are introduced. Transparency and trust are therefore key factors. This is perhaps most graphically illustrated in the US examples, where officials must persuade the public to vote for a LET proposal before it can be introduced.

The above discussion highlights that transferability is a far from simple matter. It will depend not just on whether a LET can technically and legislatively be applied, but very much on the socio-political contextual factors as well. For example the local sales taxes may be very useful and widely implemented throughout North America but this does not immediately mean that the system could easily be transferred to Europe and form a reliable source of funding for public transport. Firstly, there may not be the legislative system for such local taxation to be levied, but even if legislation were passed to permit the use of such a LET, the other socio-political aspects would remain absent.

There is one group of LETs that are increasingly used in a wide variety of cultural, legislative and social-political situations. These are road user and parking charges. In very contrasting situations, charging for the use of roads is increasingly viewed as a way of raising money for public transport while at the same time pricing the externalities of the car. This also holds, albeit to a lesser extent for parking charges and local fuel taxes. So the potential for these LETs seems to be somewhat higher, especially in the highly congested

Figure 6.1 Bus entering the Durham road user charging zone.
In this historic city centre in north-east England, congestion was adversely affecting its tourist and shopping functions. With the introduction of a road user charge in 1992, traffic was cut by 90%. Proceeds from the charge part-fund the 'Cathedral Bus' that links the charge zone to park and ride sites and the rail station.

areas where a consensus has emerged that serious action is needed to address
a serious transport problem.

A search for success factors

Given the complexity behind the success, failure, and applicability of the
various LET mechanisms, it is an intriguing research question to determine
what the critical factors are for the design of a LET. The approach taken is
to deploy a recently-developed approach for comparative case studies which
originates from multi-criteria analysis and meta-analysis. Such methods have
become an established technique for taxonomic purposes in the medical,
decision and natural sciences, especially in the case of comparative analysis of
(semi-)controlled case study experiments (see, among others, Van den Bergh
et al., 1997). At present, these methods are also extensively used in the social
sciences, especially in experimental psychology, pedagogy, sociology and,
more recently, in economics (see Matarazzo and Nijkamp, 1997). In general,
comparative case study analysis aims to synthesize previous research findings
or case studies with a view to identifying commonalties which might lend
themselves to be transferred to other, as yet unexplored cases. While there
has been significant methodological progress in quantitative case study fields,
in situations of low measurement scales (qualitative, categorical or ordinal
data), standard statistical techniques cannot be deployed. This applies to our
empirical case study data on LETs, therefore we have resorted to a recently
developed method for qualitative multi-dimensional classification analysis,
called *rough set theory* (for details, see Pawlak, 1991, Slowinski, 1995
and Van den Bergh *et al.*, 1997). This has its origins in the field of artificial
intelligence and is able to incorporate not only different measurement scales,
but also different degrees of measurement precision known as granularity in
classification experiments. Rough set analysis is therefore very appropriate for
analysing our LETs cases.

The basis of rough set analysis is formed by a categorical data matrix,
called the information matrix. Qualitative information on attributes or
performance values of case studies (objects) is systematically represented in
this information matrix, and the application of rough set analysis to this data
table then makes it possible to identify which possible combinations of values
or attributes, measured in distinct classes, are compatible with certain ranges
of performance variable. The 'decision rules' are then specified as 'if . . . then'
statements, based on qualitative information. If certain attributes have a high
frequency of occurrence in all decision rules, then this means that they tend
to exert a dominant influence on the performance indicator characterizing
the case study concerned and hence may be considered to be critical success
factors. If an attribute shows up in all decision rules, this is the core of the

impact system and may therefore be regarded as the dominant critical success factor.

For a comparative analysis of the case studies on public transport systems, the relevant characteristics or attributes which were likely to exert an impact on the success of the case study were systematically explored and assessed in order to identify critical success factors. In other words, we had to identify several criteria or characteristics that are likely to have an impact on the level or direction of success. Overall, eleven characteristic attributes were defined as being critical success factors for funding public transport, based on the available data (see also the previous section) and a sensitivity analysis. These were:

1   Approval (the attribute that refers to the level of decision making): in particular, on which level were implementation decisions made (i.e. regional, national or via voter approval);

2   Revenues: what was the amount of the hypothecated revenues raised by the schemes during a given year (i.e. three bands were used, <€9m, €9m–€17m, >€17m);

3   Principle: what is the principle behind the taxation schemes? (i.e. polluter pays, beneficiary pays or someone pays (spreading the burden));

4   Public acceptability: what did the public think about the schemes in terms of acceptance – without any objections or were there (some or many) reservations? (categorized as high, medium or low acceptability);

5   Transferability (expressed in terms of easiness of implementation somewhere else): in particular, does the scheme need many changes, e.g. by law or technology?

6   Complexity: is there sufficient simplicity for users and administrators in terms of collection of payments? (very complex, complex, simple, very simple);

7   Flexibility: does the case allow for fine tuning of the charge level and the mechanism in general, or is it difficult to implement (e.g. via a change of law)? (high, medium or low difficulty);

8   Links to transport policy: to what extent does the taxation scheme affect other modes of transport? (strong, medium, weak);

9   The level of ambition of the scheme: how ambitious are the schemes? (high, medium, low ambition). A low level of ambition corresponds to a general tax or charge, a part of which is hypothecated; highly ambitious schemes are identified as packages with clear objectives other than funding related (e.g. reducing congestion);

10 The geographical element: where is the scheme located? (Europe, United States, Asia);

11 The typology of the scheme: what type of innovative funding? (for the nine identified categories see Table 6.2).

The success of a scheme was hence the decision variable – in rough set terms – in the empirical comparison. It was thus an endogenous variable, to be explained by the characteristics of a particular scheme. The answer to the question of whether or not a case could be characterized as a success was extracted from an extensive literature search and expert opinions on the case studies concerned. Here, a success was interpreted as the achievement of a scheme's objectives. This could differ amongst the various cases, as objectives were not always the same. So, it need not be based on the amount of money generated by the scheme. A simple success score on a 4-point scale was adopted, which varied from whether a case has been very successful, moderately successful (some criticism), not really successful, or of unknown outcome. Cases that have never been implemented were placed in this last category.

Table 6.4 A classification table for the qualitative measurement of the attributes of world-wide LET public transport funding cases

Qualitative classes for attribute values

Attributes		1	2	3	4	5	6	7	8	9
A1	Approval	regional	national	local voter						
A2	Revenues (year)	<€9m	€9m –17m	>€17m	unknown					
A3	Principle	polluter	bene-ficiary	spreading the burden						
A4	Public acceptability	high	medium	low	unknown					
A5	Transferability	high	medium	low						
A6	Complexity	very complex	complex	simple	very simple					
A7	Flexibility	high	medium	low						
A8	Linked to transport policy	strong	medium	weak	none					
A9	Ambition	high	medium	low						
A10	Geographical location	Europe	United States	Asia						
A11	Scheme	Charges for use of roads	Con-sumption taxes	Local motor-ing taxes	Employ-er taxes	Property taxes	Devel-oper levies	Parking charges	Utility levies	
Success, described as the achievement of the scheme's objectives (effectiveness)		yes	yes, but some critics	no	unknown					

A success was denoted by the code 1, a moderate success by the code 2, while the code 3 was given when the scheme was not (or hardly) a success, and the code 4 for an unknown result. Table 6.4 presents the classification table for the codes.

The information table on the cases can be found in Table 6.5.[2] The furthest left vertical column presents all the schemes that have been included in the analysis. Horizontally one can find the attributes (A1 = approval, A2 = revenues etc.) and the success score. For example, the first scheme included

Table 6.5 Presentation of the information table

Location of the scheme (type)	A1	A2	A3	A4	A5	A6	A7	A8	A9	A10	A11	Success
Vienna (ET)	1	3	2	4	2	3	2	4	2	1	4	2
Paris (ET)	1	3	2	1	2	3	2	4	2	1	4	2
Portland (ET)	1	3	2	4	2	3	1	4	2	2	4	2
Vancouver (PT)	1	3	2	4	3	3	2	4	3	2	5	2
San Francisco (PT)	3	3	2	2	3	3	3	4	3	2	5	2
Los Angeles (PT)	1	3	2	1	3	4	1	4	3	2	5	2
Hamburg (DL)	1	2	2	4	1	3	1	2	1	1	6	2
San Francisco (DL)	1	1	2	3	2	3	3	3	2	1	6	2
Hong Kong (DL)	1	3	2	4	2	3	1	4	3	2	6	1
Milton Keynes (PC)	1	1	1	2	1	5	1	1	2	1	7	1
London airports (PC)	1	1	1	1	1	5	1	1	3	2	7	2
Aspen (PC)	3	1	1	2	1	5	2	1	2	1	7	3
La Spezia (PC)	1	1	1	2	1	5	1	1	3	1	7	1
Amsterdam (PC)	1	1	1	2	1	4	1	1	2	2	7	3
Oslo (CR)	1	3	1	2	2	2	2	2	2	1	1	2
Trondheim (CR)	1	2	1	3	2	2	2	3	3	1	1	3
Bergen (CR)	1	2	1	2	2	2	2	3	3	1	1	3
Singapore (CR)	2	3	1	4	2	1	2	1	1	3	1	1
San Diego (CR)	3	1	1	1	3	2	1	1	2	2	1	1
San Francisco (CR)	1	3	1	2	1	4	3	3	2	2	1	2
Hong Kong (CR)	2	3	1	3	2	2	2	1	1	3	1	4
Stockholm (CR)	2	4	1	3	2	1	2	2	1	1	1	4
Cambridge, U.K. (CR)	1	4	1	4	2	1	1	1	1	1	1	4
Florida (MT)	3	3	1	2	1	4	3	1	3	2	3	1
State of Washington (MT)	1	3	1	4	1	4	3	1	3	2	3	1
Reno (CT)	3	2	3	2	3	3	3	4	2	2	2	1
Forth Worth (CT)	3	3	3	2	3	3	3	4	2	2	2	1
Atlanta (CT)	3	3	3	2	3	3	3	4	2	2	2	1
Austin (CT)	3	1	3	2	3	3	3	4	2	2	2	1
Birmingham US (CT)	1	1	3	4	3	2	3	3	3	2	2	4
Pennsylvania (CT)	3	3	3	1	3	3	3	4	3	2	2	2
Arizona (CT)	3	1	3	1	3	3	3	4	3	2	2	2
Pullman (Wash.) (UL)	3	1	3	2	2	4	3	4	3	1	8	2
Wuppertal (Germ.) (UL)	2	4	3	1	2	3	1	4	2	2	8	1
Berkeley (MI)	3	1	3	2	2	4	3	4	2	2	9	1
JFK Airport (MI)	1	4	1	2	1	3	2	2	2	2	9	2

Notes:
ET = employer tax, PT = property tax, DL = developer levies, PC = parking charges, CR = charges for the use of roads, MT = local motoring taxes, CT = consumption taxes, UL = utility levies, MI = miscellaneous.

in the analysis is the one from Vienna. This is an employer tax scheme which has been regionally approved and subsidizing public transport for more than €17m. The scheme has been moderately successful, as indicated by the success score of 2.

The data of Tables 6.4 and 6.5 were used as input for the software tool ROSE 2. ROSE is a modular software system implementing basic elements of the rough set theory and rule discovery techniques (Predki and Wilk, 1999). The output from the experience consisted *inter alia* of 'minimal sets'. These minimal sets can be described as deterministic conditions, under which certain attributes show up in the performance measure of all cases. Of course, given the combinatorial nature of the rough set methodology, many minimal sets may emerge. Hence, it was of particular interest to identify those attributes that show up in all minimal sets (referred to as the 'core'). This is because attributes in the core may be regarded as critical conditions for the performance of the LET cases. From the empirical analysis, there appeared to be thirteen minimal sets (see Table 6.6). With these sets it was, in principle, possible to explain the success or failure of a certain LET case. However, this does not mean that for each minimal set it is always possible to determine a meaningful effect relationship, which can be interpreted on substantive grounds. Nevertheless, these minimal sets are in principle able to identify which combinations of attributes may logically – i.e. in the light of the underlying database – lead to a certain outcome of the success variable.

Table 6.6 Minimal sets of attributes 1–11 and core attributes

Minimal sets		Core
{1 2 4 6 11}	{1 4 6 8 11}	4: public acceptability
{2 4 6 7 11}	{4 6 7 8 11}	11: type of scheme
{2 4 6 10 11}	{4 6 8 10 11}	
{1 4 9 11}	{3 4 9 10 11}	
{4 5 9 10 11}	{4 6 9 10 11}	
{2 4 9 11}	{4 7 9 11}	
{4 8 9 10 11}		

Table 6.6 also shows the core of the analysis. Cores are factors that show up in all minimal sets and thus have a common explanatory value in all statements on success conditions.

In the LET analysis, two factors appeared to play a crucial role. These were public acceptability and type of scheme, which were the core variables and were (perhaps unsurprisingly) thus indispensable in explaining the success rates of the cases involved. Without these, the difference in success of a case could not be fully explained.

The other variables in the minimal sets have a lower frequency of occurrence (see Table 6.7). High frequency rates mean that these attributes stand out in a more pronounced way in the interpretation of the success rate of the LET funding cases. It was clear that, after the two core variables, the next most frequent variables were the complexity of the scheme and the level of ambition. These featured in 54% of the minimal sets. It was also noteworthy that transferability and the principle behind the schemes seemed to have very little effect on success.

These results are illuminating, as they show the relative contribution of various background factors of LET schemes on the performance of these schemes. Moreover, these results are plausible and are verified by the fact that the core and other high frequency factors are also those frequently mentioned in the literature and in personal interviews.

Table 6.7 Frequencies of attributes in minimal sets

Independent variable	Appearance in minimal sets
1. Approval (A1)	3 (23%)
2. Revenues (year) (A2)	4 (30%)
3. Principle (A3)	1 (8%)
4. Public acceptability (A4)	13 (100%)
5. Transferability (A5)	1 (8%)
6. Complexity (A6)	7 (54%)
7. Flexibility (A7)	3 (23%)
8. Linked to transport policy (A8)	4 (30%)
9. Ambition (A9)	7 (54%)
10. Geographical location (A10)	6 (46%)
11. Scheme (A11)	13 (100%)

It is also revealing to present and analyse the decision rules resulting from our rough set analysis (see Table 6.8). These rules determine the combinations of attributes that are responsible for the level of success. For example, it appears that the highest rate of success (i.e. code 1) can be explained by seven different decision rules. The entries of Table 6.8 can be interpreted as follows, taking the fourth line as an illustration. If A1 = 2 (i.e. if the approval (attribute 1) of the scheme is national (qualitative class 2)) and if A11 = 8 (i.e. if there is cross-utility financing), then there is a clear case of success. It also appears that local motoring tax schemes (A11 = 3) always have a high rate of success. In the same way, we may derive that funding public transport by property taxes (A11 = 5) leads to a minor success, as well as all employer tax cases (A11 = 4). Finally, it appears that there are only two rules that explain no success (rules 16 and 17).

Concluding remarks

Most of the categories of LETs provide a relatively stable, dedicated funding source with a high level of practicality. For many of the schemes identified, LET funding forms a substantial share of the operating budget for public transport. From our overview, it is clear that earmarked taxes have been

Table 6.8: Decision rules for performance of public transport schemes (A1 to A11 again refer to the various attributes)

Rule	Decision Variable	Decision Rule
1	1	A9 = 2 & A11 = 2
2	1	A4 = 2 & A6 = 5 & A7 = 1
3	1	A11 = 3
4	1	A1 = 2 & A11 = 8
5	1	A2 = 4 & A6 = 2
6	1	A4 = 4 & A10 = 3
7	1	A1 = 3 & A11 = 9
8	2	A11 = 4
9	2	A1 = 3 & A3 = 3 & A4 = 1
10	2	A5 = 1 & A9 = 2 & A10 = 2
11	2	A4 = 3 & A11 = 6
12	2	A11 = 5
13	2	A1 = 1 & A8 = 2
14	2	A9 = 2 & A11 = 8
15	2	A4 = 1 & A11 = 7
16	3	A3 = 1 & A7 = 2 & A9 = 3
17	3	A6 = 4 & A11 = 7
18	4	A2 = 4 & A6 = 1
19	4	A6 = 2 & A10 = 3
20	4	A6 = 2 & A11 = 2

widely implemented in the United States. In Europe, relatively few examples have been found, but some (e.g. The *Versement* in France) are significant. This pattern of using LETs is mainly due to the institutional organization in the various countries and the local character of the cases. In the United States the high use of LETs seems largely due to the combination of State level revenue-raising powers and a need for new funds to support public transport. Local authorities, responsible for the provision of public transport, have a stronger incentive or need to seek new ways of financing public transport if conventional sources of finance are limited or come under pressure. If existing sources of finance are forthcoming from central governments then there has been less need for new funding techniques, which is largely the case in Europe.

In general, LET mechanisms have evolved because 'traditional ways' of funding public transport have been cut or are viewed as politically problematic. Governments have become sensitive to the levels of general taxation, and funding for public transport is particularly vulnerable to this attitude. This is because persistent expenditure is needed over a period of time and, importantly, the results of such spending are not usually apparent within the lifetime of a single government.

There is a second and, in the context of modern transport and environmental policies, a more important reason to fund public transport via hypothecated charging. This is that these schemes are not only a means of raising financial support for public transport systems, but also as a method of sending appropriate pricing signals to transport users, with the possibility of their being integrated with more traditional general fiscal and regulatory instruments. However, the majority of existing LET measures have evolved without reference to the guiding principles of public finance. Most have been developed simply in order to generate funds to support public transport. A major question then is: how successful have these schemes been?

The rough set analysis provided some interesting results. It appears that the type of scheme and the level of public acceptability have the highest impact on the level of success. In general, the degree of success is not dependent on the transferability and principles behind the schemes but on broader considerations such as the scheme's specific features and the public support of the scheme concerned, although the principles of the scheme do exert an influence on public acceptance. For example if congestion, pollution and accidents are widely perceived problems, then this enhances public acceptance. The links between key factors, therefore, need to be understood.

The rough set results seem to offer plausible and interesting results. This method may be seen as an approach to generate common inferences from case studies on a similar subject. In this regard, rough set analysis is different from standard evaluation tools such as cost-benefit analysis, cost-effectiveness analysis, or multi-criteria analysis. The aim of these methods is to identify or to select the best possible choice alternative from a set of competing alternatives. In our approach we try to identify commonalities in order to draw general lessons from individual cases. Both classes of method are complementary and do contribute to a better understanding of complex transport policy issues (see also Banister *et al.*, 2000).

Overall, lessons from this analysis for the design of future LET mechanisms include:

- Using LET mechanisms to fund a specific project for which the need is widely accepted is likely to increase the key issues of acceptance and transparency.
- The schemes need to be as simple as possible. Complexity tends to increase costs and reduce transparency.
- It may be necessary to reduce other taxes to compensate the biggest losers from the introduction of a LET to raise acceptability. For example, reduction in fuel duty compensated by more targeted LET mechanisms or a cut in other employee taxes might be examples.
- There is a value in phased introduction of LET charges, with the flexibility to fine tune and adopt the mechanism over time. It is presently impossible to model the impacts and success of demand management

transport policy measures. Flexibility in mechanisms thus plays a key-role.

• The success of a scheme is very much dependent on local or regional circumstances. Road pricing, for example, is a clear example where success (in terms of practicality, acceptability and potential) depends on local circumstances. So, schemes identified in this analysis as successful might not be successful in other circumstances.

Notes

1   A more detailed overview of the assessment can be found in Van den Branden *et al.* (2000).
2   This information table draws on data from Van den Branden *et al.* (2000).

Chapter 7

The future of LETs

The inevitability of subsidy

Mobility is one of the essential values of modern urban civilization, but the transport needs of individuals are no longer met under acceptable conditions. The constant growth in motor-vehicle traffic causes all sorts of nuisances – air pollution, noise, accidents, and congestion – which are worsening and which people are finding increasingly difficult to bear. These negative environmental effects are not limited to major cities and big conurbations, and the contribution of transport to the emission of greenhouse gases, due mainly to exhaust fumes from motorcars, is rapidly increasing worldwide. The growth in car use and the increasing distances travelled leads to environmental deterioration, despite a constant improvement in the fuel consumption and ecological performance of cars. The dependence on the car is becoming widespread, the average distance covered is increasing, and in many areas walking, cycling and public transport are used less and less as a consequence.

Policy-makers are trying (and have tried for many years) to curb these trends. It has generally been recognized that the sustainability of the transport system is served with a modal shift from the private car to more environmentally friendly modes. Public transport is one of the alternatives to the use of private vehicles. Although public transport provides a relatively small proportion of total travel in many places, it can help address various transport problems, for instance, more efficient mobility in congested areas. Public transport requires relatively little road and parking space per passenger, offers mobility to people who cannot own or drive a car, and can help achieve energy and emission reductions and improve liveability. Due to these benefits public transport has become an interesting policy target for the improvement in the functioning of the transport system. This has led to governments supporting public transport financially.

These benefits of public transport are very much related to the theoretical justifications for policy intervention in this sector. Economic theory suggests two principal reasons for the regulation of markets by the public sector: market failure and income distribution. Market failures comprise, *inter alia*, the existence of externalities, public goods, and natural monopoly. Regulatory intervention can also be rationalized as a way to affect income distribution and promote the mobility of disadvantaged groups. It is mainly the alleged impact on negative externalities from public transport, support of favourable land-use patterns, and assistance to low-income and specific groups of users, that explain present transit regulatory and subsidy policies (Berechman, 1993).

Subsidies can have different forms. Operational subsidies may cover financial deficits totally or partly, whereas capital subsidies are just lump sum contributions from local or national public bodies. Indirect subsidies could be provided by specific infrastructure, equipment, or tax exemptions. Although justifiable, ideally public transport should not require subsidy as it should be treated as a viable, commercial industry. But direct subsidies and indirect subsidies abound in all economies, affecting the competitive position of most goods and services, including transport. So, public transport has come to rely heavily on public funds. Figures show that subsidies of 70% of the operating costs are not unusual. With a few exceptions, public transport in every country requires some degree of external financial support.

These subsidies consist of large sums of money. It is, however, increasingly difficult for governments to find sources to fund all their expenses. Governments and local authorities, faced with increasing demands on their resources, are under growing pressure to obtain good value from public transport undertakings. Given that the general budget is not inexhaustible anymore, privatization, franchising, and private finance initiatives are increasingly common and in some countries, such as Britain, certain categories of support have been reduced, or even abolished. Meanwhile a less common response has been for cash-strapped public transport authorities to explore new sources of funding, of which several can be classed as earmarked taxes and changes. It is these that have been exmained in this book.

LETs: an overview

Local earmarked taxes and charges (LETs) have emerged over a number of years and in different specific situations. We have found a wide range of examples all over the world whereby local taxation is used to fund public transport. All of them have been reported in the previous chapters. It is therefore not surprising that the design of these various LETs measures has placed different emphasis on the principles and purposes of taxation considered above. There appear to be three main groupings, depending upon who pays:

- Beneficiary Pays (to pay for collective goods and services)
- Polluter Pays (as an instrument of environmental/transport policy in order to discourage social harm) and

♦ Spreading the Burden (to achieve a broad base of taxation).

Certain aspects of public transport have a collective consumption benefit. These include the economic benefits to towns and cities and labour force benefits to employers. The Beneficiary Pays principle seeks to use LETs to charge people and organizations for these collective benefits. This could involve a local charge to the area where public transport investment takes place. Examples include LETs on employment, on property, on land values and, on developers. The French *Versement Transport* is probably the best known example of making the beneficiary pay, while the developer levy is most common, with payments from developers being required to compensate for the transport impacts generated.

The 'polluter pays' principle is a well-known concept, which is at the heart of European Union transport policy. Pollution charges are fees or taxes imposed on polluters based on the amount of pollution. Transport pollutes in a variety of ways and causes many external effects for which some are charged in certain parts of the world. LETs of this type appear to be the fiscal instrument that fits most closely with modern transport and environment policy principles. Examples include parking fees, road user charges, and local fuel excise duties. Their potential seems to be high due to increased political attention and the high rate of social acceptance of spending the revenues from transport taxes within the transport sector (and improving the substitutes for car transport). The trend towards more innovative measures of this kind is therefore clearly understandable.

For the remainder of LETs, notably the majority of those used across the United States, the major principle behind adopting particular revenue sources has been to raise as much money in as low profile and uncontroversial way as possible. The guiding principle is that of 'spreading the burden'. These measures are essentially designed to provide a broad tax base. Beyond that, questions of whether such taxes are fair, equitable or contribute to meeting environmental goals do not seem to have featured in any significant way. For example, the general sales tax, the most widely used earmarked charge in the United States for funding public transport, is regressive. It falls disproportionately on the poorer in society and does not discourage social harm.

Chapter 6 showed that the fund raising potential appears to be high for most categories of LETs, and they can often form a substantial part of the operating budget or contribute in a significant way to the construction of new infrastructure. On the issue of practicality, most of the examples rely on existing legislative structures, which keep costs and complexity relatively low. Only some of the more recent LETs (e.g. road pricing) have required new powers. This tendency to rely on existing legislative structures has implications for flexibility and transferability. It may result in the 'easy' being implemented rather than what is appropriate. Furthermore, with many LETs being implemented under regional or local-specific powers, transferability can be limited. Even if circumstances and institutional aspects are often similar, it appears that it is not that simple to copy successful examples. It should not be forgotten that certain categories (e.g. local motoring taxes and consumption taxes) are very much a product of conditions and taxation systems prevailing in the United States. Existing institutional structures (organizing referenda) and generally low tax levels make it easier to implement new taxes to fund public transport.

A further point is that the market and policy environments of Europe and North America are very different. The United States is very much a car-oriented society and public transport use is low, which makes it difficult to raise support for additional funding for a service that is hardly used except by the poor. By contrast, in Europe the lack of such processes and existing structures (taxation on a national level), as well as the already relatively high taxes on fuel would make the transfer of US style mechanisms very difficult.

The issue of public acceptability is not an easy one to address. Often, acceptability is low when a new charge or tax is introduced, but improves when the objective (i.e. to fund public transport) is explained and the new or improved services are introduced. Transparency and trust are therefore key factors. This is perhaps most prominently illustrated in the US examples, where officials must persuade the public to vote for a LET proposal before it can be introduced.

In summary, in order to get more feeling for factors determining the level of success, we deployed a new classification tool, the *rough set analysis*. It appears that the type of scheme and the level of public acceptability have a high impact on the level of success. In general, the degree of success is not dependent on the transferability and principles behind a scheme, but on broader considerations such as the scheme's specific features and the public support for it. However, the principles of the scheme do exert an influence on public acceptance. For example, if congestion, pollution, and accidents are widely perceived and accepted problems, this enhances public acceptance. The links between key factors therefore, need to be better understood.

LETs and public policy

Although in recent years the use of LETs for local transport demand management has attracted growing attention, in general the more established LETs were developed simply as a source of income to support and/or enhance public transport services or to fund their expansion. This is linked to two trends in public transport finance and provision.

The first relates to difficulties with the traditional forms of financing public transport investment, i.e. grants to municipalities from central government. There is a great deal of competition for public money, and transport often falls behind spending demands for education and health. This has led to a search for new sources of funding, which have included the private sector (via privatization or contracting agreements) and earmarked LETs.

The second factor is that there has been a trend in a number of EU Member States, and elsewhere, to devolve the responsibility for local and regional public transport away from national government. This has led to the desire to devolve funding mechanisms too, which in some cases has involved the development of LETs. This could be viewed as an example of subsidiarity in action.

We have seen that in countries with a federal system, the local state or region usually has some form of local taxation within its control. In such cases, the introduction of LETs need not involve any new funding legislation, but can be accommodated within existing structures. The United States is an obvious example of such a situation, where in response to a reduction in Federal support for public transport, individual States and cities used their powers to raise a whole variety of local taxes in order to support and develop public transport systems. Regional transport authorities have also been formed to deal with suburban-urban transport issues. But with public transport little used in the United States, it is difficult to communicate the necessity of new taxes (to the public) keeping unused services alive.

The combined pressures of limits to traditional funding sources and devolution of responsibility for public transport, on their own or combined, have resulted in a gradual development of LETs as a source of public transport funding. In general, this development has not been linked to any particular theory or concept of public finance, although in Europe the principle of subsidiarity has played some role. When introducing such new funding mechanisms, the local situation is of key importance. Fundamentally, the policy-maker must choose the mechanism that most effectively delivers the required objectives within the political, institutional, and financial constraints.

Until recently these objectives have tended to focus on increasing public transport use for social and environmental reasons, while the constraints were centred on the reluctance of politicians to raise taxes too obviously or risk upsetting car users. However, in Europe at least the ground is shifting. Here, it is now realized that mobility management, i.e. restraining car use, is a desirable policy objective in itself – rather than a constraint as before. As a consequence LETs have been advocated from a different perspective and by a different set of actors with transport and environment ministries now seeing LETs as a tool of transport demand management. This has particularly applied to measures such as road user charging, area pricing, congestion charging, parking charges, and development levies. Although fund raising is far from ignored, the basis of such developments has been entirely separate from existing LETs. Furthermore the traffic control and charging element has resulted in such 'new' LETs being subject to additional political controversy.

LETs: key lessons

This book has shown that LETs have a potentially powerful role as an instrument addressing twenty-first century transport and environment policy needs. It is clear that the use of LETs has some clear advantages in a world that is looking for external financial support. In particular LETs can be:

♦ well targeted and use related, providing both carrots and sticks;
♦ locally empowering; subsidiarity of financing as well as decision-making. This also improves acceptability of charging mechanisms because regional paid taxes are used for improvements within the region;
♦ stable and reliable sources of finance.

On the other hand, LETs do also have some limitations:

♦ they are not the best instrument for every situation;
♦ the easy money syndrome;
♦ are often used to avoid the real decisions (e.g. needing to reform management). This is a weakness of subsidies in general.

Evaluating the experiences with LETs highlights various issues. Some are in line with the polluter pays principle, which has an impeccable logic (charging polluters for the damage they do). But charging beneficiaries for the advantages they gain may be psychologically more attractive, and therefore easier to accept politically.

Earmarking and efficiency is another issue. The combined system of the way funds are raised and how they are spent is crucial in this context. Standard public finance theory argues that revenues should be spent on activities yielding the highest social return rather than being set aside for a dedicated purpose. An alternative is that the public should be able to choose the charging and expenditure packages they prefer (as happens in the United States with proposals for local levies). However, research into the cost effectiveness of public transport expenditure in Paris indicates good cost-effectiveness of expenditure with no efficiency losses. So, hypothecation does not appear to lead to inefficiency in itself. This review shows that there is a strong case that some degree of hypothecation is important in gaining public acceptance and accountability for economic instruments, and that they do not yield major problems of inflexibility and efficiency.

Overall, this analysis for the design of future LET mechanisms suggests:

- using LET mechanisms to fund a specific project for which the need is widely accepted is likely to increase the key issues of acceptance and transparency;
- the majority of existing LETs have evolved without reference to guiding principles of public finance. Most have been developed simply in order to generate funds and have no direct relationship to the principles of the polluters pay or fair and efficient pricing;
- the schemes need to be as simple as possible. Complexity tends to increase costs and reduce transparency;
- there is a value in phasing the introduction of LET charges, with the flexibility to fine tune and adopt the mechanism over time. It is presently impossible to model the impacts and success of demand management transport policy measures. Flexibility in mechanisms thus plays a key-role;
- the success of a scheme is very much dependent on local or regional circumstances.

LETs tend to offer new solutions where traditional policies may fail. It ought to be recognized that the general public finance foundations often fall short when it comes to public support and implementation. Public transport is not an undisputed good; it is often criticized because of inefficiency, low level of service, or bureaucratic attitude and often does not command a high level of sympathy among all citizens. In a policy situation where many stakeholders compete for scarce financial resources, public transport finds itself in an uneasy position. It may well have to find its main financial revenues from conventional funding schemes, but any increase or diversity in funding in a creative way has to be welcomed. Therefore, hypothecation may be a critical means to generate extra support.

Clearly, there is no single, simple and unambigious system of LETs that can be adopted by local authorities everywhere. There is a great diversity, which demonstrates how creative local decision-makers are in developing support initiatives in cases where the traditional finance paradigms fail. This also means that such paradigms have to be extended with principles from transport demand management theory, industrial organization, urban and regional planning, green taxation schemes, or generalized road user charging principles. In theory, one might even think of a collective fund of hypothecated revenues from a variety of LETs that might create a more sound financial basis

– due to its flexibility, local orientation, and local targeting – for an upgrading (or even 'quality jump') in local public transport provision. Such a new situation would of course prompt the need for a more focused public finance system theory based on a blend of financial pooling and fiscal federalism. It seems that from both a theoretical and policy viewpoint, exciting challenges lie ahead.

AC Transit (1999) UC Berkeley –AC Transit Class Pass, General Manager's News Briefs, 8 May. Visit http://www.actransit.org/. Last accessed June 2002.

Ahm, K. (1999) Personal communication. Head of Finance, Ørestadsselskabet, Copenhagen.

APTA (American Public Transport Association) (1999) APTA Transit Statistics 1997. Visit http://www.apta.com. Last accessed December 2002.

Arizona State Lottery (2002) *Where the Money Goes*. Arizona State Lottery. Visit http://www.arizonalottery.com/money_goes/default.asp. Last accessed 28 May 2002.

Banister, D., Stead, D., Steen, P., Akerman, J., Dreborg, K., Nijkamp, P. and Schleicker-Tappeser, R. (2000) *European Transport Policy and Sustainable Mobility*. London: Spon Press.

Belk, R. (1995) Acknowledging consumption, in Miller, D. (ed.) *Studies in New Consumer Behaviour*. London: Routledge.

Bell, D.D. (1993) *Funding Methods for Urban Rail Construction and Improvements in Japan*. Public Transit: Current Research in Planning, Marketing, Operations and Technology. Transportation Research Record No 1401. Washington DC: Transportation Research Board, National Research Council, pp. 9–16.

Berechman, J. (1993) *Public Transit Economics and Deregulation Policy*. Studies in Regional Science and Urban Economics No 23. Amsterdam: North Holland.

Birmingham Jefferson County Transit Authority (2002) *Chronology*. Visit http://www.bjcta.org/about_chronology.html. Last accessed 22 May 2002.

Bjerrgaard, R., Bangemann, M., and Papoutis, C. (1996) Auto Oil Programme Document, COM(96) 248 Final, Sheet 23. European Commission, DGXII, Brussels.

Black, A. (1995) *Urban Mass Transportation Planning*. London: McGraw Hill.

Brown, G. (2001) Personal communication. Western Australia Department of Transport, August.

Bushell, C. (1994) *Jane's Urban Transport Systems 1993/94*. Coulsdon, Surrey: Jane's Information Group.

Button, K. and Rietveld, P. (1993) Financing urban transport projects in Europe. *Transportation*, **20**, pp. 251–265.

Cale, L. and Almond, L. (1992) Physical activity levels in young children: a review of the evidence. *Health Education Journal*, **51**(2), pp. 94–99.

Casino Revenue Fund Advisory Commission (2002) 2001 Casino Revenue Fund: Helping New Jersey Senior Citizens and Persons with Disabilities. Leaflet, New Jersey Casino Control Commission, Department of the Treasury, Office of Management & Budget, Trenton NJ. Visit http://www.state.nj.us/casinos/crf.pdf. Last accessed 28 May 2002.

Castells, M. (1996) *The Rise of the Network Society*. Oxford: Oxford University Press.

CEC (Commission of the European Communities) (1995) *Towards Fair and Efficient Pricing in Transport. Policy options for internalising the external costs of transport in the European Union*. Green paper. COM (95) 691. Brussels: CEC.

CEC (Commission of the European Communities) (1996) *The Citizens Network*. Luxembourg: Bureau for Official Publications of the EU.

CEC (Commission of the European Communities) (1998) *Fair Payment for Infrastructure Use: A phased approach to a common infrastructure charging in framework in the EU*. White Paper. Brussels: CEC.

CEC (Commission of the European Communities) (2001) *European Transport Policy for 2010: Time to Decide*. White Paper. Brussels: CEC.

Cervero. R. (1983) Views on transit financing in the US. *Transportation*, **12**(1), pp. 21–43.

Chartered Institute of Public Finance and Accountancy (1974) *Passenger Transport Operations*. London: CIPFA.

Chen, H.W. and Fang, S.H. (1998) *Air Pollution Control Fee: The Taiwan Experience*. Visit http://www.epa.gov.tw/english/offices/f/fee.htm. Last accessed 18 July 2001.

City of Louisville (2000) Budget 2000. Mayor of Louisville Kentucky. http://www.louky.org/mayor/budget2000. Last accessed 20 June 2001.

City of Pullman (2000) Discussion of Levels of Services and Priorities: Transit, Report to the City Council. Visit http://www.ci.pullman.wa.us/transitlevelsofservice.htm. Last accessed 31 May 2002.

Coindet, J.P. (1994) Financing urban public transport in France, in Farrell, S. (ed.) *Financing Transport Infrastructure*. London: PTRC.

Collins, S. (2002) Personal communication. Cambridgeshire County Council, Cambridge, July.

Commission for Integrated Transport (2001) *European Best Practice in Delivering Integrated Transport*. London: CfIT.

Commission on Taxation and Citizenship (2000) *Paying for Progress: A new politics of tax for public spending*. London: Fabian Society.

Copenhagen Transport, Danish Ministry of Transport, Arhus sporveje (1995) *The Financing of Urban Public Transportation Systems*. Draft Report, Copenhagen.

CUTR (Center for Urban Transport Research) (1996) Lessons Learned in Transit Efficiencies, Revenue Generation, and Cost Reduction, Report by the Center for Urban Transport Research, College of Engineering, University of Southern Florida.

Dalvi, M.Q. and Patankar, P.G. (1999) Financing a metro rail through private sector initiative: the Mumbai Metro. *Transport Reviews*, **19**(2), pp. 141–156.

DETR (Department of the Environment, Transport and the Regions) (1998*a*) *A New Deal for Transport – Better for Everyone*. London: The Stationery Office.

DETR (Department of the Environment, Transport and the Regions) (1998*b*) *Breaking the Logjam*. London: The Stationery Office.

Department of Transport (1989) *National Road Traffic Forecast (Great Britain)*. London: HMSO.

Deran, E.Y. (1965) Earmarking and expenditures: A survey and national test. *National Tax Journal*, December, pp. 354–361.

Dutch Ministry of Housing, Physical Planning and Environment (1990) *National Environmental Policy Plan + (NEPP+)*. Den Haag: SDU Publishers.

EC (European Commission) (1997) Communication from the Commission to the Parliament, the Economic Social Committee and the Committee of the Regions, Climate Change – The EU approach for Kyoto. COM(97)481 final, October.

EC (European Commission) (1999) *Panorama of Transport: Statistical overview of road, rail and inland waterway transport in the European Union 1970–1996*. Eurostat, Theme 7 Transport. Luxembourg: Office for Official Publications of the European Communities.

ECMT (European Conference of Ministers of Transport) (1998) *Efficient Transport in Europe: Policies in internalisation of external costs*. Paris: OECD/ECMT.

EFTE (European Federation for Transport and the Environment) (1994) *Green Urban Transport: A survey*. Preliminary report 94/2. Brussels: EFTE.

Eggler, B. (2001) Canal Streetcar project officially is on track: work on new line to begin in August. *The Times-Picayune* (New Orleans), 21 July.

Elsom, D. (1996) *Smog Alert: Managing urban air quality*. London: Earthscan.

Else, P.K. (1992) Criteria for local transport subsidies, *Transport Reviews*, **12**, pp. 291–309.

Environmental Protection Agency of Taiwan (1998) Taiwanese Pollution Fee Additional. Visit http://www.epa.gov.tw/english/offices/f/bluesky/bluesky9.htm. Last accessed 18 July 2001.

European Parliament (1991) *Economic and Fiscal Instruments of Environment Policy*. Report of the Committee on the Environment, Public Health and Consumer Protection, European Parliament, Report A3-0130/91, DOC_EN\RR\109943. Strasbourg: European Parliament.

Eurostat (1997) *EU Transport in Figures – Statistical pocketbook*, 2nd issue. Luxembourg: European Union.

Farrell S. (1999*a*) *Financing European Transport Infrastructure: Policies and practice in Western Europe*. Basingstoke: Macmillan.

Farrell S. (1999*b*) Personal communication. Imperial College, University of London, 27 July.

Feltz, H. (1992) Organisation and Financing of Public Transport in West Germany. Paper presented to UITP Conference, Budapest.

Florida Department of Transportation (2000) *Florida's Transportation Tax Sources: A Primer*. Tallahassee, Florida: Office of Management and Budget, Florida Department of Transportation.

Freund, P. and Martin, G. (1993) *The Ecology of the Automobile*. Montreal: Black Rose Books.

Gemeente Amsterdam (1999) *Begroting 1999*. Amsterdam: Gemeente Amsterdam.

GGBHTD (Golden Gate Bridge, Highway and Transportation District) (1999) The Golden Gate Bridge Highway and Transportation District Homepage. Visit http://www.goldengate.org. Last accessed June 2003.

GGBHTD (Golden Gate Bridge Highway and Transportation District) (2000*a*) *Highlights, Facts and Figures*, 5th Edition. San Francisco, CA: Golden Gate Bridge Highway and Transportation District.

GGBHTD (Golden Gate Bridge Highway and Transportation District) (2000*b*) Annual Report 1999–2000. San Francisco, CA: Golden Gate Bridge Highway and Transportation District.

Gibbons (2000) Personal communication. Metro Council, Minneapolis/St Paul, Minnesota, January.

Goldman, T., Corbett, S. and Wachs, M. (2001) Local Option Transportation Taxes in the United States. Research Report UCB-ITS-RR-2001-3I. Institute of Transportation Studies, University of California, Berkeley.

Gómez-Ibáñez, J.A. (1999) Pricing, in Gómez-Ibáñez, J. A., Tye, W. B. and Winston, C. (eds.) *Essays in Transportation Economics and Policy*. Washington: Brookings Institution Press, pp. 99–136.

Goodwin, P.B. (1994) Traffic Growth and the Dynamics of Sustainable Transport Policies. Transport Studies Unit, University of Oxford.

Goodwin, P.B. *et al.* (1991) Transport: The New Realism. Report to the Rees Jeffreys Road Fund. London, March.

Gourd (1999) Personal communication. City of Hamburg Building Department, Hamburg, Germany.

Gubbins, E.J. (2002) *Managing Transport Operations*. London: Kogan Page.

GVTA (Greater Vancouver Transportation Authority) (1999) *1999 Operating and Capital Budget, Transport 1999*. Vancouver, British Columbia: Greater Vancouver Transportation Authority.

Hass-Klau, C. and Crampton, G. (1999) How other countries see light rail. *Tramways and Urban Transit*, March, pp.100–102.

Hillman, M. (ed.) (1993) *Children, Transport, and the Quality of Life*. London: Policy Studies Institute.

Houghton, J.Y., Jenkins, G.J. and Ephraums, J.J. (1990) *Climate Change: the IPCC scientific assessment*. Cambridge: Cambridge University Press.

Ieromonachou P., Enoch, M.P. and Potter, S. (2003) All charged up: early lessons from the Durham Congestion-Charging Scheme. *Town and Country Planning*, **72**(2), pp. 44–48.

IEA (International Energy Agency) (2001) *The Road from Kyoto: Current CO_2 and transport policies in the IEA*. Paris: OECD/IEA.

Jensen, J. (2002) Personal interview. Land Valuation Assessment Officer, City of Copenhagen, June.

Jones, J. (1999) Personal communication. Manager of Financial Forecasting, Tri-Met. Portland, Oregon.

Joubarne, L. (2001) Personal communication. Agence Metropolitaine de Transport, Montreal, November.

Kay, J.H. (1997) *Asphalt Nation*. New York: Crown Publishers.

Knox, J.K. (1996) *Benefit Assessment Districts. Enhancing the quality of life in California*. Sacramento, California: Planning & Conservation League and PCL Foundation. Visit http://www.pcl,org/Store/benereport.html. Last accessed 29 May 2002.

Kramhöller (1999) Personal communication. Magistrat der Stadt Wien, Vienna, 12 August.

Lamb, J. (1999) Personal communication. Airport Access Manager, BAA Stansted. London.

LTC (Legislative Transportation Committee) (2001) *Transportation Resource Manual*. Olympia, Washington: Washington State Legislature. Visit http://ltc.leg.wa.gov/Manual01/toc.htm. Last accessed 31 May 2002.

Litman, T. (2002) Evaluating Public Transit Benefits and Costs. Victoria Transport Policy Institute, British Columbia. Visit http://www.vtpi.org.

Lyons, G. and Chatterjee, K. (2002) *Transport Lessons from the Fuel Tax Protests of 2000*. Aldershot: Ashgate Press.

Marshall, S., Banister, D. and McLellan, A. (1997) A strategic assessment of travel trends and travel reduction strategies. *Innovation, The European Journal of Social Sciences*, **10**(3), pp. 289–304.

Marshall, S. and Banister, D. (2000) Travel reduction strategies: intentions and outcomes. *Transportation Research*, Part A, **34**, pp. 321–338.

Matarazzo, B. and Nijkamp, P. (1997) Methodological complexity in the use of meta-analysis for empirical environmental case studies. *International Journal of Social Economics*, **34**(719), pp. 799–811.

Mathie, I. (2001) Personal communication. City of Edinburgh Council, June.

Metro Council (1999) *Public Transit – Metro Transit*. Minneapolis/St Paul, Minnesota: Metro Council. Visit http://www.metrocouncil.org/transit/metrotra.htm. Last accessed June 2002.

Metropolitan Transportation Authority of Los Angeles (2002) Benefit Assessment Districts Program, Benefit Assessment Districts Program Office, MTA Los Angeles. Visit http://www.mta.net/trans_planning/CPD/BAD/default.htm. Last accessed 29 May 2002.

Meyer, A. (1996) *Le Versement Transport*. Report to the Union des Transport Public, Paris.

Ministère de l'Aménagement du Territoire, de l'Equipement et des Transports (1995) *Urban Public Transport in France: Institutional Organisation*. Paris: Land Transport Administration, Ministère de l'Aménagement du Territoire, de l'Equipement et des Transports.

Ministry of Provincial Revenue (1993) Motor Vehicle Parking: Social Service Tax Act. Consumer Taxation Branch, Revenue Programs Division, Ministry of Provincial Revenue, Bulletin 105. Victoria, British Columbia. Visit http://www.rev.gov.bc.ca/ctb/publications/bulletins/105.htm. Last accessed July 2002.

Ministry of Transport (1963) *Traffic in Towns: A study of the long term problems of traffic in urban areas*. Reports of the Steering Group and Working Group appointed by the Minister of Transport. London: HMSO.

Mitoula, R., Patargias, P. and Abeliotis, K. (2003) The Transformation Efforts in the City of Athens (Greece) towards Environmentally Friendly Transportation. Paper presented at the 39th International Planning Congress (ISOCARP), Cairo, 17–22 October.

Mohring, H. (1972) Optimization and scale economies in urban bus transportation. *American Economic Review*, **62**, pp. 591–604.

Moore, N. (2001) Personal communication. Team Leader Implementation, Planning Division of Bracknell Forest Council, Berkshire.

Nakagawa, D. and Matsunaka, R. (1997) *Funding Transport Systems: A comparison among developed countries*. Oxford: Pergamon.

Nash, C.A. (1988) Rail investment: The continental perspective. Institute for Transport Studies, University of Leeds.

National Transit Database (2001) *National Transit Database Profiles 2000.* Washington DC National Transit Database, Federal Transit Administration, Department of Transport,. Visit http://www.ntdprogram.com. Last accessed 31 May 2002.

Nelson Nygaard Consulting Associates (2001) *Technical Impact Development Fee Analysis.* San Francisco: Nelson Nygaard Consulting Associates.

OECD (1995) *Motor Vehicle Pollution: Reduction strategies beyond 2010.* Paris: OECD.

OECD (1999) *Indicators for the Integration of Environmental Concerns into Transport Policies, Environment Directorate.* Paris: OECD.

Oregon Department of Transportation (2001) Program Budget 2001–2003 Biennium. Governor's Recommended Budget. Oregon Department of Transportation.

Oregon State Legislature (1999*a*) Cigarette and Tobacco Products Tax, Oregon Revised Statutes, Chapter 323, 1999 Edition. Visit http://landru.leg.state.or.us/. Last accessed 1 November 2001.

Oregon State Legislature (1999*b*) Mass Transportation, Oregon Revised Statutes, Chapter 391, 1999 Edition. Visit http://landru.leg.state.or.us/. Last accessed 1 November 2001.

Parking Review (2002) Special Parking Areas are spreading across the country. *Parking Review*, June, pp.14–15.

Patrikalakis, Y. (1999) Personal communication. OASA, Athens, Greece, 10 August.

Pattison, A. (1999) *Jane's Urban Transport Systems.* Coulsdon, Surrey: Jane's Information Group.

Pattison, A. (2001) *Jane's Urban Transport Systems, 2001–2002.* Coulsdon, Surrey: Jane's Information Group.

Pawlak, Z. (1991) *Rough Sets.* Dordrecht: Kluwer.

Pennsylvania State Lottery (2002) Where does the money go? Visit http://www.palottery.com. Last accessed 28 May 2002.

Pessaro, B. (2001) Personal communication. SANDAG, San Diego, California.

Potter, S. (2000) Travelling Light. Theme 2, Course T172, Technology Level 1, Working with our environment: Technology for a sustainable future. The Open University, Milton Keynes.

Potter, S., Enoch, M. P., Rye, T. and Black, C. (2001) *The Potential for Further Changes to the Personal Taxation Regime to Encourage Modal Shift.* Final Report to the Department of the Environment, Transport and the Regions, June. Visit http://www.dtlr.gov.uk/itwp/modalshift/index.htm. Last accessed June 2003.

Predki, B. and Wilk, S. (1999) Rough set based data exploration using ROSE system, in Ras, R.W. and Skowron, A. (ed.) *Foundations of Intelligent Systems, Lecture Notes in Artificial Intelligence*, No. 1609. Berlin: Springer-Verlag, pp. 605–608.

Proost, S. and van Dender, K (including partners) (1999) *Trenen II Stran.* Final Report. Catholic University of Leuvan, Belgium.

Pucher, J. (1988) Urban public transport subsidies in Western Europe and North America. *Transportation Quarterly*, **42**(3), pp. 377–402.

Pucher, J. (1999) Transportation trends, problems, and policies: an international perspective. *Transportation Research Part A*, **33**, pp. 493–503.

Pucher, J. and Lefèvre, C. (1996) *The Urban Transport Crisis in Europe and North America*. Basingstoke: Macmillan.

Reggiani, A. (1999) Personal communication. Bologna University, Bologna, Italy, September.

Rice Center (1986) *Alternative Financing for Urban Transportation: the state of the practice*. Final Report. Houston: Rice Center.

Ridley, T.M. and Fawkner, J. (1987) Benefit Sharing: the funding of urban transport through contributions from external beneficiaries. Paper presented to the 47th International Conference, International Railways Committee, UITP, Lausanne.

Rivenburg, K. (1999) Personal communication. Lane County Mass Transit District, Eugene, Oregon.

Rothengatter, W. (2001) Transport subsidies, in Button, K.J. and Hensher, D.A. (eds.) *Handbook of Transport Systems and Traffic Control*. Oxford: Pergamon.

Rye, T. (2002) *Palace Quarter, Den Bosch Case Study. Results, conclusions and recommendations*. Integrating Mobility Management in Spatial Planning (OPTIMUM) CD-Rom. Brussels: EU Interreg IIc Programme.

San Diego State University Foundation, Jacqueline Golob Associates, Bob Maxwell & Associates, Resource Decision Consultants, and Schreffler E (1997) *Worldwide Experience with Congestion Pricing*, I-15 Congestion Pricing Project Monitoring and Evaluation Services, Task 1.1. Prepared for the San Diego Association of Governments, 10 June. Visit http://www.sandag.com. Last accessed 1 July 2001.

Schumacher, D. (2001) Personal communication. Metropolitan Transit Development Board, San Diego, California.

Serageldin, I. (1993) Environmentally Sustainable Urban Transport: Defining a Global Policy. Paper presented to the fiftieth International Congress of the Union Internationale des Transport, Sydney, Australia.

Shreffler, E. (2001) Personal communication. Independent consultant, San Diego, California.

Simpson, B.J. (1994) *Urban Public Transport Today*. London: E&FN Spon.

Sims, L. and Berry, J. (1999) Various Ways of recovering Increases in Land and Property Values – The example of North America. Urban Public Transport Funding, UITP Seminar, Paris, 13–14 October.

Slack, E. (2001) Alternative Approaches to Taxing Land and Real Property. Research Paper, Enid Slack Consulting Inc., October. Visit http://www.worldbank.org. Last accessed 31 May 2002.

Slowinski, R. (1995), *Intelligent Decision Support: Hardbook of Applications and Advances of Rough Set Theory*. Dordrecht: Kluwer.

Small, K.A. and Gómez-Ibáñez, J.A. (1999) Urban transportation, in Cheshire, P. and Mills, E.S. (eds.) *Handbook of Regional and Urban Economics*, Volume 3: *Applied Urban Economics*. Amsterdam: North-Holland.

SORTA (Southwest Ohio Regional Transit Authority (undated) *Metro/SOTA Financial Information*. Cincinnati, Ohio: SORTA. Visit http://www.sorta.com/ff-budget.ssi. Last accessed 20 June 2001.

Sound Transit (2002) Sound Transit's 2001 revenues exceed forecast budget despite recession. Sound Transit, Seattle, Washington, March 21. Visit http://www.soundtransit.org. Last accessed 31 May 2002.

TBTA (Triborough Bridge and Tunnel Authority) (2002) Welcome to MTA Bridges and Tunnels, Metropolitan Transit Authority, New York City, New York. Visit http://www.mta.nyc.ny.us/badt/html/btintro.htm. Last accessed 31 May 2002.

TCRP (Transit Cooperative Research Program) (1998) *Funding Strategies for Public Transportation*. TCRP Report 31, v1/2, Project H-7 FY 95, Transportation Research Board, National Research Council, Federal Transit Administration. Washington DC: National Academy Press.

Teitz, M. (1999) Fees, taxes and planning. *Town & Country Planning*, 68(4), pp.140–141.

Teja, R.S. and Bracewell-Milnes, B. (1991) *The Case for Earmarked Taxes: Government spending and public choice*. Research Monograph 46. London: Institute of Economic Affairs.

Thoms, P. (2001) Personal communication. New South Wales Department of Transport, June

TransLink (2002*a*) *2002 Annual Report*. Vancouver, British Columbia: Greater Vancouver Transportation Authority. Visit http://www.translink.bc.ca. Last accessed January 2003.

TransLink (2002*b*) It's time to choose. Leaflet. Greater Vancouver Transportation Authority, Vancouver, British Columbia.

Transport for London (2000) *International Fares Comparison: London, Paris, New York*. London: Integration Department, Transport for London.

Transport Road Research Laboratory (1980) *The Demand for Public Transport*. Crowthorne, Berkshire: TRRL.

Tri-Met (2001) Facts about Tri-Met: Transit Works. Leaflet. Tri-Met, Portland, Oregon. Visit http://www.tri-met.org. Last accessed 31 May 2002.

Tsukada, S. and Kuranami, C. (1994) Value capture: the Japanese experience, in Farrell, S. (ed.) *Financing Transport Infrastructure*. London: PTRC.

University of Washington Transportation Office (2000) *Moving People: U-Pass Annual Report 1999–2000*. Seattle: U-Pass Program, Transportation Office, University of Washington. Visit http://www.washington.edu/upass/. Last accessed July 2001.

USEPA (US Environmental Protection Agency) (1999) *A Guidebook of Financial Tools – Paying for sustainable environmental systems*. Washington DC: Office of the Chief Financial Officer, US Environmental Protection Agency. Visit http://www.epa.gov/efinpage/guidebook/guidebookp.htm. Last accessed 22 May 2002.

Usher, L. (1998) Marketing transportation alternatives to visitors and locals: How the City of Aspen makes a difference in traffic congestion without using tax payers money. *Proceedings of the Association of Commuter Transport*. ACT Annual Conference, San Francisco, 30 August–2 September. Washington DC: Association for Commuter Transport.

Vaccare M A (1996) *An Overview of Innovative Financing: Highway and Transit*. Transportation Research Record No. 1527, Transportation Law Issues, Transportation Research Board. Washington DC: National Academy Press, pp. 31–34.

Van den Bergh, J.C.J.M., Button, K., Nijkamp, P. and Pepping, G. (1997), *Meta-analysis in Environmental Economics*. Dordrecht: Kluwer.

Van den Branden, T., Ubbels, B., Enoch, M.P., Potter, S., Nijkamp, P., and Knight, P. (2000) *Fair and Efficient Pricing in Transport – The role of charges and taxes*. Final Report for European Commission DG TREN. Birmingham: Oscar Faber. Published on: http://www.epommweb.org/links_frame.html.

Van Doren, C. (1992) *A History of Knowledge*. New York: Balantine Books.

Von Weizsäcker, E. and Jesinghaus, J. (1992) *Ecological Taxation Reform*. London: Zed Books.

Waerstad, K. (2002) Personal communication. Norwegian Public Roads Administration, July.

Washington State Department of Revenue (2002) *Tax Reference Manual: Information on State and Local Taxes in Washington State*. Olympia, Washington: Research Division, Washington State Department of Revenue. Visit: http://dor.wa.gov/docs/reports/2002/Tax_Reference_2002/contents.htm. Last accessed July 2002.

Washington State Department of Transport (2000) *2000 Summary of Public Transportation Systems in Washington State*. Olympia, Washington:

Washington State Department of Transport. Visit: http://www.wsdot.wa.gov/transit/library/2000_summary_links.cfm. Last accessed July 2002.

Weir, T. (1999) Personal communication. Transport Department, City of Aspen, Colorado, 28 May.

Whelan, J. (2003) *Funding London's Transport Needs*. London: Royal Institute of Chartered Surveyors.

Whitelegg, J. (1992) Ecological taxation reform, in Whitelegg, J. (ed.) *Traffic Congestion: is there a way out?* Hawes: Leading Edge, pp. 169–183.

WHO Centre for Environment and Health (1995) *Residential Noise: Concern for Europe's tomorrow.* Stuttgart: Wissenschaftliche.

Wilson, M. (2001) Personal communication. Dublin Transportation Office, Dublin, Ireland.

Wong, K., Lim, L.C. and Chan, S.H. (2002) Personal interview. Land Transport Administration, Singapore, 5 July.

XE (2001) Currency Converter, XE.com. Visit http://www.xe.com. Last accessed 31 May 2002.

Zahavi, Y. (1973) The travel-time relationship: A unified approach to transportation planning. *Traffic Engineering & Control*, August/September, pp. 205–212.